T0184734

"The relationship between climate change and tourism has for too long been under-researched in the southern African context. This extremely welcome volume makes a timely contribution to highlighting the potential repercussions of climate change and its management for the economically significant yet also extremely vulnerable tourism and hospitality sectors. Drawing on wider themes of resilience and sustainability, as well as issues of mitigation and adaptation, this book will provide a valuable benchmark for African climate change and tourism education and research."

— Professor C. Michael Hall, Department of
Management, Marketing and Entrepreneurship,
University of Canterbury

"Tourism has emerged as an increasingly important contributor to the economy of southern African countries and, despite the enormous setbacks brought about by the COVID-19 pandemic, remains a significant development factor going forward. The relationships between tourism and the environment - and climate in particular - are multi-dimensional and complex, such that a deeper understanding of their dynamics is needed - especially in the era of anthropogenic climate change. The authors here present a convincing and comprehensive analysis of the challenges to the tourism industry brought about by a rapidly changing climate and, in so doing, offer a timely and authoritative perspective on an issue that has considerable relevance to the achievement of the Sustainable Development Goals in the region."

— Professor Michael E. Meadows, President of the
International Geographical Union

Climate Change and Tourism in Southern Africa

This book explores the nature of climate change in southern Africa, its impacts on tourism, and the resilience, adaptation, and governance needs in various tourism operations and environments. Previous studies on climate change and tourism have mainly focused on the Global North and specific forms of tourism such as snow-based winter activities. Drawing on case studies from a wide range of countries including South Africa, Lesotho, Namibia, Botswana, and Zimbabwe, this book fills this lacuna by describing and analysing the climate change and tourism nexus in the southern African context. The book begins by providing an overview of the current and estimated impacts of climate change to the tourism industry in the region, highlighting the deepening socio-economic inequities, and environmental and social injustices. It focuses on the importance of sustainable tourism in tackling these issues and highlights that resilience and robust governance and policy systems are essential for a tourism destination to successfully adapt to change. By synthesising the key lessons learned through this analysis, *Climate Change and Tourism in Southern Africa* also draws attention to specific adaptation and policy strategies which have value for other regions in the Global South. This book will be of great interest to students and scholars of climate change, tourism, and environmental policy and justice.

Jarkko Saarinen is professor of Human Geography at the University of Oulu, Finland, and Distinguished Visiting Professor (Sustainability Management) at the University of Johannesburg, South Africa.

Jennifer Fitchett is associate professor of Physical Geography at the University of the Witwatersrand, South Africa.

Gijsbert Hoogendoorn is professor of Geography at the University of Johannesburg, South Africa.

Routledge Advances in Climate Change Research

For more information about this series, please visit: www.routledge.com/ Routledge-Advances-in-Climate-Change-Research/book-series/RACCR

Climate Change and Tourism in Southern Africa

Jarkko Saarinen,
Jennifer Fitchett and
Gijsbert Hoogendoorn

Routledge
Taylor & Francis Group

LONDON AND NEW YORK

from Routledge

First published 2022
by Routledge
4 Park Square, Milton Park, Abingdon, Oxon OX14 4RN

and by Routledge
605 Third Avenue, New York, NY 10158

Routledge is an imprint of the Taylor & Francis Group, an informa business

British Library Cataloguing-in-Publication Data
A catalogue record for this book is available from the British Library

Library of Congress Cataloging-in-Publication Data
A catalog record has been requested for this book

ISBN: 978-0-367-56750-7 (hbk)
ISBN: 978-0-367-60942-9 (pbk)
ISBN: 978-1-003-10261-8 (ebk)

DOI: 10.4324/9781003102618

Typeset in Times New Roman
by codeMantra

Contents

Illustrations

List of boxes

List of figures

Box figures

List of tables

Box contributors

Julius Atlhopheng is a professor of environmental science at the University of Botswana. He has served as Head of Department and Dean of Faculty of Science. He completed his MSc at Kings College London (University of London, UK) and PhD at University of Wollongong (Australia). His research interests are on dryland science, climate change, and environmental sustainability management.

Kayla Mac Conachie is currently enrolled for an MSc in Geography at the University of the Witwatersrand, reconstructing the rainfall patterns of the Namib Desert through tree rings.

Kiran Dookhony-Ramphul is a lecturer of tourism management at the University of Mauritius. Her research interests are sustainable tourism development, entrepreneurship, and tourism SMEs. She has driven the development of the first undergraduate BSc and Diploma in tourism programmes at the Faculty of Law & Management where she teaches.

Kaitano Dube is an associate professor of ecotourism and National Research Foundation of South Africa rated researcher at the Department of Hospitality, Tourism and Public Relations Management, Faculty of Human Sciences, at the Vaal University of Technology (VUT). He is well published in areas of tourism, climate change, aviation, and tourism sustainability.

Francois Engelbrecht is a distinguished professor of climatology at the Global Change Institute of the University of the Witwatersrand. He is one of the lead authors of the Intergovernmental Panel on Climate Change (IPCC).

Anne Fitchett is an associate professor of project management in the School of Civil and Environmental Engineering and the assistant dean undergraduate in the Faculty of Engineering and the Built Environment at the University of the Witwatersrand.

Jonathan Friedrich holds an MSc in Geography from Georg-August University towards which he completed a research project exploring the threats of climate change to beach tourism in South Africa. He is currently a doctoral researcher at the Leibniz Centre for Agricultural Landscape Research.

Wame L. Hambira is a senior research scholar at the Okavango Research Institute of the University of Botswana. Her research interests include tourism and climate change adaptation and tourism-dependent communities' resilience to climate change. She holds a PhD in Geography from the University of Oulu, Finland and an MSc in Environmental Economics, University of York (UK).

David Lesolle is a lecturer in environmental science with specialisation in climate science, applied climatology at the University of Botswana. His research interest and publications include sustainable development, climate issues, and policy. He has a wealth of experience in climate policy and has participated in various intergovernmental processes on sustainable development and climate change.

Fazlin McPherson holds an MSc in Geography from the University of the Witwatersrand, investigating the climate change threats to heritage tourism in the Makgabeng region. She is currently employed at the University of Johannesburg as a senior academic advisor.

Naomi N. Moswete is a senior lecturer in the Department of Environmental Science, University of Botswana. Her research interests include human geography, tourism as a strategy for rural development, community-based tourism, transboundary conservation areas – ecotourism nexus, parks – people relationships, heritage management, and cultural tourism. She is an editor in CABI Tourism Cases, sectional editor in *Parks and Recreation Administration Journal.*

William Mushawemhuka holds a PhD in Geography from the University of Johannesburg, exploring the climate change threats to nature-based tourism in Zimbabwe. He is currently a lecturer in the Department of Geography, Environmental Management and Energy Studies at the University of Johannesburg.

Kirsten Noome holds an MSc in Geography from the University of the Witwatersrand, presenting Tourism Climate Index scores calculated for Namibia. She currently works at CityRock.

Micheal T Pillay holds an MSc in Geography from the University of the Witwatersrand, investigating tropical cyclone dynamics across the Southern Hemisphere. He is currently enrolled for a PhD at Nagasaki University.

Ariel Prinsloo holds an MSc in Geography from the University of the Witwatersrand, exploring the impacts of the 'Day Zero' drought on the Airbnb operators in Cape Town. She is currently employed as a research officer at WWF South Africa.

Bradley Rink is an associate professor of human geography at the University of the Western Cape and the editor of the journal *Urban Forum*. His research interests include mobilities, urban place-making, and tourism geographies.

Sarah Roffe holds a PhD in Geography from the University of the Witwatersrand, exploring the classification of South Africa's rainfall zones. She is currently employed as a postdoctoral researcher in the Evolutionary Studies Institute at the University of the Witwatersrand.

Jarkko Saarinen is a professor of human geography in the University of Oulu, Finland, and distinguished visiting professor (sustainability management) at the University of Johannesburg, South Africa.

Ngoni Courage Shereni is a PhD student in the School of Tourism and Hospitality at the University of Johannesburg in South Africa and a Faculty member in the Department of Accounting and Finance at Lupane State University in Zimbabwe. His research interests include sustainable tourism, community-based tourism (CBT), and community-based natural resource management (CBNRM).

Tamzyn Smith is currently enrolled for an MSc in Geography at the University of the Witwatersrand, exploring the vulnerability of the South African Tourism sector. For her Honours research report, she explored the impacts of drought on nature-based tourism at Sabi Sands Wildtuin.

Jannik Stahl holds an MSc in Geography from Georg-August University, towards which he completed a research project exploring the threats of climate change to beach tourism in South Africa.

Ian Steyn holds an MSc in Geography from the University of the Witwatersrand, exploring the incidence of extreme heat events in Namibia. He currently works at PWC.

Lara Stockigt holds an MA in Geography from the University of Johannesburg, investigating the threats of climate change to snow tourism in Lesotho. She is currently employed as a lecturer in the Department of Geography, Environmental Management and Energy Studies at the University of Johannesburg.

Preface and acknowledgements

Climatic conditions frame and provide the backdrop to some of the most fundamental questions in tourism studies and management. The climate provides an (at least partial) explanation as to why certain places attract tourists, why people want to take a break from their everyday living environment, and why some places have flourished with tourism while others have failed. However, the patterns and processes characterising the climatic conditions of many tourism destinations have become increasingly unpredictable and changing. When writing this book, July 2021 was recorded as the world's hottest month ever measured during 142 years of global recording history. Globally, the previous record of the hottest month ever in 2016 was paralleled in 2019 and 2020. Moreover, the highest temperature recordings have intensified in the past years in different parts of the world, including southern Africa. Indeed, the climate has become warmer and increasingly characterised by unpredictable extreme events that can take place in almost any given location in the world. In southern Africa, warming temperatures have been identified as exceeding the global mean rate and are often coupled with extreme drought events. Throughout 2020 and 2021, towns and cities along the Eastern Cape coast of South Africa have suffered with chronic water shortages, while the Cape Town 'Day Zero' drought of 2015–2017 was more widely reported. On the opposite end of the scale, climate change is also heightening the frequency of severe tropical cyclones and extending their trajectories poleward. Cyclone Idai in March 2020 and Cyclone Eloise in January 2021, both of which made landfall on the city of Beira, provide insight into the devastation associated with these intense storm events.

In August 2021, the Intergovernmental Panel on Climate Change (IPCC) released the Sixth Assessment Report on Climate Change. The United Nations Secretary-General António Guterres described the findings as a code red for humanity. Indeed, it has become evident that we cannot keep our heads in the proverbial sand – we need to act to both mitigate and adapt to climate change. Since the First IPCC Assessment Report (AR1) in 1990, we have been firmly and continuously informed about the importance of climate change, the related threats, and the mitigation needs as key challenges for the future of humanity. While climate change has often been perceived as a slow process which gradually materialised in the future, the IPCC report Global Warming of 1.5°C in 2019 indicated that we have only 12 years to save the Earth as the home of humankind. This was argued to require a binding global agreement with supporting actions to significantly reduce carbon dioxide (CO_2) emissions. Now, we have less than ten years to do that.

This book aims to synthesise the existing knowledge and encourage conversation on climate change, focusing on the threats and impacts in the southern African tourism context. As a region, southern Africa is vulnerable to the impacts of climate change as it is exposed to various climate risks including heatwaves, extreme droughts, storms, and sea level rise (SLR). The region also experiences ongoing challenges of resource availability and a weak adaptive capacity. Peripheral and rural areas and communities in particular are highly vulnerable because of poverty and other social, livelihood, and financial constraints. Concerning the regional tourism industry, climate change will have far-reaching consequences that are complex, interwoven, and still partly unknown. In general, the sector and its increasing CO_2 emissions contribute to global climate change, but at the same time, tourism is highly dependent – both directly and indirectly – on favourable and predictable climatic conditions. The core attraction elements of the southern African tourism sector rely on natural landscapes, wilderness, and wildlife; these are increasingly threatened by the local and regional impacts of global climate change. Furthermore, a large segment of the region's international tourism is based on long-haul flights. The economic and social viability of long-haul flights may be negatively impacted by international climate change policies, taxation, and responsible consumer trends in future. All this makes the southern African tourism industry very vulnerable to climate change, both directly and indirectly. Thus, it is obvious that climate change signifies various threats to tourism and its sustainability and resiliency in the region.

In this book, we acknowledge the interdependent and conflicting relationships between tourism and climate change in the southern African context. Tourism is highly dependent on climatic resources, but it also contributes negatively to climate change by its operations and emissions. At the same time, however, we realise that tourism has the potential to provide regional and local development and well-being that are highly needed in southern Africa. Our grounding argument is that this calls for responsibility and sustainability in tourism development and the creation of supporting governance and policy structures for managing tourism and climate change relationships in the region and beyond. Without this, all the high-level talks and strategies on the industry's capacity to contribute to the United Nations Sustainable Development Goals (SDGs), for example, and local community and environmental well-being are just talk, hiding the past growth-oriented realities and prospects of the global tourism industry.

To have this possibility to write about tourism and climate change relations in southern Africa, we owe a great deal to many people and institutions.

First, we would like to express our gratitude to the publisher, and in particular Annabelle Harris as our editor and the editorial assistant Jyotsna Gurung, for supporting us all the way. Furthermore, we want to thank collectively and warmly the colleagues who contributed their time and expertise by writing illustrative case example boxes to the different chapters of the book: great thanks! On his behalf, Jarkko would like to thank the co-authors for the fruitful collaboration and the University of Botswana and colleagues Julius Atlhopheng and Haretsebe Manwa for initially making it possible to become familiar with tourism and environmental change in the region by hosting him as a visiting professor and later by creating an opportunity to be inaugurated to the new professorship of tourism management at the University. Jarkko would also like to thank Wame Hambira, Lenao Monkgogi, and Naomi Moswete at the University of Botswana, Chris and Jayne Rogerson, Diane Abrahams, Tembi Tichaawa, and Vyasha Harilal at the University of Johannesburg, Thandi Nzama at the University of Zululand, Gustav Visser at the Stellenbosch University, Maano Ramutsindela at the University of Cape Town, Ellen Kimaro at the University of Namibia, C. Michael Hall at the University of Canterbury, and Robin Nunkoo at the University of Mauritius. Lagotto Romagnolo and Villa Pablo & sauna have also played their positive role in writing and in-between writing. Finally, great thanks to the University of Johannesburg for providing a status position that has kept him busy with the southern African academic collaboration!

Jennifer offers great thanks and acknowledgement to the many students who have joined her over the years in studying the climate change – tourism nexus: Bronwyn Grant, Dean Robinson, Ariel Prinsloo, William Mushawemhuka, Lara Stockigt, Su-Marie Fortune, Kirsten Noome, Mazozo Mahlangu, Charné Jordan, Devania Govender, Subhashinidevi Pillay, Fazlin McPherson, and Tamzyn Smith. Thanks to the Faculty of Science of the University of the Witwatersrand for granting a sabbatical which allowed for focused time to be spent in writing this monograph. A great thanks, albeit strangely place as he is a co-author, to Gijsbert Hoogendoorn for bringing me a pile of pages of 'squiggly lines' (translate: mathematical equations) eight years ago – a move that kickstarted an incredible near-decade of work on tourism and climate change in southern Africa. Thank you to Anne and Meg for reading and critiquing the first drafts of most things I ever write. Although not in the tourism – climate change research field, a huge thanks to the support, friendship, and mentorship from Anson Mackay (UCL) and Chris Curtis (UJ) over the years.

Gijsbert would like to first and foremost thank Fatima Vally for everything (literally). A very special word of thanks to Jennifer Fitchett for explaining what the 'squiggly lines' meant and for forming a very productive collaboration, standing with me through thick and thin, the highs and the lows. Thanks to Jarkko Saarinen (University of Oulu/University of Johannesburg) for guiding us through this process and many other things too. Thanks to the highly influential people of my career thus far: Gustav Visser (Stellenbosch University), the second home gang Dieter Müller, Roger Marjavaara, Andreas Back (Umeå University), Clare Kelso and William Mushawemhuka (University of Johannesburg), Etienne Nel (Otago University), Daniel Hammett (The University of Sheffield), Irma Booyens (University of Strathclyde), and Lochner Marais (University of the Free State). Thanks to all the post-graduate students who worked with me through the years especially Jannik Stahl and Jonathan Friedrich (The Georg August University of Göttingen). Lastly, to Pieter de Wet for being my guitar teacher at 16 and my philosophy guide at 40.

Jarkko, Jennifer, and Gijsbert

1 Introduction

Introduction

The Sixth Assessment Report (AR6) of the Intergovernmental Panel on Climate Change (IPCC) states unequivocally that human-induced global environmental change severely affects earth's biota, with consequences on physical, biological, and human systems (IPCC, 2021). The tourism sector is contributing to this problem through heavy use of resources, greenhouse gas emissions, pollution, and pressure on infrastructure and the natural environment at large (Gössling & Scott, 2018). The tourism sector is also particularly vulnerable to the impacts of global environmental change, with rising sea levels, droughts, floods, storms, reduction in snow, and heatwaves already impacting destinations across the globe (Scott et al., 2016b). The significance of these vast impacts of global environmental change on the tourism sector, and the tourism sector on global environmental change, has prompted the academic community to realign scientific inquiry towards these urgently pressing issues (Becken, 2013).

The tourism and climate change nexus has become an important and voluminous field of investigation in Tourism Geography, Tourism Studies, Hospitality, Leisure, and Tourism Management (Fang et al., 2018). While tourism and climate change research dates back to the early 1990s (see Smith, 1990; Wall & Badke, 1994), the vast majority of research has been published over the past decade. It comes as no surprise that most of the scientific publications and highly cited work has emanated from the Global North, specifically from Europe, North America, and Australasia (Becken & Hay, 2007; Hall & Higham, 2005; Scott et al., 2012). Within these varied geographical locations, a range of themes have emerged within broader topics including adaptation, mitigation, and sustainability, with focal areas such as marine tourism, snow tourism, and mountainous environments and using varied tools of analysis (Kaján et al., 2015; Peeters & Dubois, 2010; Scott, 2021; Scott et al., 2006).

The Global South has seen significantly less investigation despite being arguably more vulnerable to the impacts of climate change and global environmental change more broadly due to limited adaptive capacities

DOI: 10.4324/9781003102618-1

(Bossio et al., 2019; Haddad, 2005). The cases that have been investigated are spatially very localised. For example, the country that has seen the most research attention within the Global South is China, the global economic powerhouse and source market for many destinations globally (Fang & Yin, 2015; Lin & Matzarakis, 2011). Research attention in Latin America and Africa is particularly low and geographically specific (Dilimino & Dickinson, 2015; Navarro-Drazich, 2019; Rutty et al., 2021). Despite the general lack of research in the Global South, and particularly in Africa (Hambira & Mbaiwa, 2021; Hoogendoorn & Fitchett, 2018a), the southern African region nevertheless has become productive research locale with a large body of research on tourism and climate change emerging from the region since the mid-2000s (Hoogendoorn & Fitchett, 2020).

The purpose of this book is to consolidate the research on tourism and climate change in southern Africa to establish a coherent narrative on the key themes that have emerged over the past decade and a half. This will allow the development of strong new research foci for future investigation and will assist a variety of stakeholders in the tourism industry, government, and non-governmental organisations to develop an informed perspective on the influence of climate change on the tourism sector in southern Africa. Furthermore, we aim to demonstrate that although Rutty et al. (2021, p. 641) argue that "major geographical gaps persist in Africa" in climate, tourism, and recreation research, this is not the case for the southern African scholarship in this domain. Indeed, we demonstrate that research on climate change and tourism in the region is active and novel, characterised by international perspectives with context sensitive approaches and research questions.

An abridged history of the southern African region

When considering contemporary tourism in southern Africa, it is important to reflect on the colonial and post-colonial context of the region. The southern African region has a rich cultural and political history, originating from the pre-colonial times (Barnard, 2019; McKenna, 2010; Naidu, 2008). Starting in the 17th century, the southern African region made up a significant part of the colonial empires of Europe such as the Dutch and British in South Africa (over different time periods), the British protectorates currently referred to as Botswana, Lesotho, Eswatini, and British settlement in Zimbabwe (Adhikari, 2010). In addition, German settlement took place in Namibia (later to be occupied by South Africa), and the Portuguese took control of large areas of what is today Mozambique during the scramble for Africa (Dedering, 2000; O'Laughlin, 2002). During different time periods, the Dutch, British, and French controlled and occupied the South Indian Ocean Islands of Madagascar, Comoros, Reunion, and Mauritius (Allen, 2001).

The fight for and achievement of independence from the colonial yoke occurred at different times periods throughout the 20th century. In general terms, however, many southern African countries achieved independence a lot later than countries in west, east, and North Africa (Meredith, 2006). The first country in the region to gain independence was South Africa in 1910, but only attaining an inclusive democratic dispensation following the fall of Apartheid in 1994 (Hyam & Henshaw, 2003). In 1990, the Apartheid South African government ceded South-West Africa, which became Namibia (Christopher, 1988; Saunders, 2009). Botswana gained independence in 1960, Lesotho in 1966, and Eswatini (then Swaziland) in 1968. Zimbabwe gained independence in 1980 after 15 years of armed struggle known as the Rhodesian Bush War (Onslow, 2011). After drawn out war for independence, Mozambique established independence in 1975 (Hoogendoorn & Back, 2019).

Unfortunately, the post-independence and post-colonial periods in southern Africa have been marred by political instability, civil war, and a lack of economic growth (Nel, 2003). At present, there is a relative political stability in the region, with some success stories such as Botswana, but economic growth has remained sluggish, especially from the region's economic centre South Africa, that since the 2008 global economic downturn has not recovered fully (Fedderke, 2018). However, one key example of economic success in growth in the region has been tourism (Hoogendoorn, 2021), with international flows that still largely follow patterns from the colonial period and visitors from the former metropoles.

A brief introduction to tourism in southern Africa

For the purpose of this book, the southern African region is limited to include Botswana, Eswatini, Lesotho, Mozambique, Namibia, South Africa, Zimbabwe, Madagascar, Mauritius, and Reunion. The regional classification is not an officially recognised economic or geographic boundary, and many studies in the region may define the boundaries or inclusion criteria differently (Hoogendoorn & Rogerson, 2016; Rogerson & Rogerson, 2011; Saarinen et al., 2009). Southern Africa boasts a wide variety of destinations, many leveraging nature-based tourism (NBT), beach tourism, and urban tourism, with strong sub-sectors such as domestic, inter-regional, visiting friends and relatives (VFR) travel, and second home tourism (Rogerson & Hoogendoorn, 2014; Visser & Hoogendoorn, 2011). Some of the most well-known tourism destinations are the Okavango Delta in Botswana (Mbaiwa, 2005), Sossusvlei in Namibia (Lapeyre, 2011), the Victoria Falls in Zimbabwe (Dube & Nhamo, 2019a), the Kruger National Park (Ferreira & Harmse, 2014) and Table Mountain (George, 2010) in South Africa, Hlane Royal National Park in Eswatini (Lukhele & Mearns, 2013), Katse Dam and Afriski in Lesotho (Noome & Fitchett, 2019), and the World Heritage Site *Ilha de Moçambique* in Mozambique (Dantas & Mather, 2011). South Africa is the dominant tourism market in the region with 15,825,296 tourist arrivals in 2019 (StatsSA, 2019). The most recent statistics for Botswana reflect 1,830,274 tourist arrivals in 2018 (Statistics Botswana, 2020) and for the same year in Namibia, 1,557,279 arrivals (Ministry of Environment and Tourism, 2018). Also, for 2018, Zimbabwe recorded 2,579,974 arrivals (Zimbabwe Tourism Authority, 2018), Lesotho recorded 1,173,000 arrivals (Hoogendoorn et al., 2021), and Mozambique recorded 2,743,000 arrivals (Trading Economics, 2020). In 2019, Eswatini recorded 792,408 tourist arrivals (Eswatini Tourism Statistics, 2019).

The COVID-19 pandemic that spread across the globe during the first quarter of 2020 has significantly changed the landscape of tourism (Nhamo et al., 2020) and has unsurprisingly crippled and in some cases destroyed tourism operations in southern Africa (Rogerson & Baum, 2020). As a long-haul destination for Europe, North America, and emerging markets such as India and China (Visser & Hoogendoorn, 2011), the long-term effects of the pandemic for southern Africa will be felt for years to come (Rogerson & Rogerson, 2020). Economic recovery is possible in due course; however, the impact of current and future climate change will fundamentally change the tourism sector globally and will have major long-term impacts. Therefore, it is important to keep the long-term consequences of climate change in mind rather that the relatively short-term impacts of COVID-19.

The southern African tourism sector is particularly vulnerable to climate and other stressors due to a variety of co-occurring factors. The first and most critical factor is the extreme lack of adaptive capacity (Hoogendoorn & Fitchett, 2018b). Classified as (lower or upper) middle-income countries, Botswana, Eswatini, Mauritius, Namibia, and South Africa, and as low-income countries in the case of Mozambique and Zimbabwe, immediate needs of the population take greater precedence over investment in long-term adaptation for future climate hazards (Chiutsi & Saarinen, 2017; Vincent, 2007). A large proportion of tourist attractions in southern Africa are outdoors, which heightens the reliance on good weather conditions (Giddy et al., 2017a; Pandy & Rogerson, 2019). Many southern African tourism businesses are classified as Small Medium and Micro Enterprises (SMME), including guest houses, bed-and-breakfasts, boutique hotels, and attractions such as guided tours, safari excursions, and small museums or craft markets (Booyens & Visser, 2010). The influence of multi-national corporations on the tourism sector has increased in the region since the 1990s, but remains minimal by global standards (Rogerson & Visser, 2007). These small tourism businesses seldom have the financial capital and backing to survive major environmental shocks or to invest in infrastructure to withstand low-probability but high-intensity climatic events and are therefore much more vulnerable than, for example, major hotel chains (Fitchett et al., 2016c). Second homes tourism is a common form of domestic tourism in the region (Hoogendoorn, 2011a), and here, the common single-owner nature of second home ownership in southern Africa means that severe impacts such as floods, storms, and droughts could force owners to abandon or liquidate their properties (Hoogendoorn & Fitchett, 2018b).

Much of southern Africa has large rural populations (Ritchie & Roser, 2018). With low levels of urbanisation and dispersed and low-density populations across very large geographical areas, such as in Namibia, Botswana, and Mozambique, impoverished governments are often thinly spread when providing support to both rural and urban communities that rely on tourism (Hambira & Saarinen, 2015; Shackleton et al., 2015). Therefore, understanding community resilience and community livelihoods remains a core need in research on climate change and tourism in southern African region (Kaján & Saarinen, 2013). The influence of climate change will not exist in isolation; southern African will continue to face a number of challenges impacting livelihoods across the region.

A brief bibliometric overview of climate change and tourism literature in southern Africa

As of September 2020, at least 57 publications on climate change and tourism in southern Africa were listed between the *Scopus* and *Google Scholar* databases (Figure 1.1). The first academic publication on the intersection between climate change and tourism in southern Africa was a book chapter written by Preston-Whyte and Watson in 2005 (Hoogendoorn & Fitchett, 2020). The chapter titled '*Nature Tourism and Climatic Change in Southern Africa*' was published in Hall and Higham's (2005) seminal text *Tourism, Recreation and Climate Change*. This chapter laid the foundation for future research in the southern African context, yet little further work empirical research until Hambira's (2011) work on the Okavango Delta in Botswana (Figure 1.1). Following lower productivity during 2013 and 2014, from 2015, the study of tourism and climate change in southern Africa has seen uninterrupted publication (Figure 1.1). The year 2020 has been thus far the most productive year in southern African tourism and climate change research, with 16 publications.

Although the subject of some of the earliest research in the region, research in Botswana has been far surpassed by the bulk of literature from South Africa (Figure 1.2). In 2016 and 2017, all the publications produced came from South Africa, and in 2018, locations expanded to Namibia, Lesotho, and Zimbabwe. In 2019, most research occurred in South Africa, but with a strong increase of publications on locations in Zimbabwe and some publications from Mauritius and Lesotho. During 2020, Zimbabwe became a focal study location in the region. Within South Africa, Cape Town has been the focus of greatest number of studies, with nine publications. Cape Town is southern Africa's main long-haul tourist destination and offers a variety of tourism products such as beach tourism, NBT, wine tourism, urban tourism, and heritage tourism (Frey & George, 2010). Other key geographical areas of study are the predominantly urban tourism destination of Gauteng Province with the cities of Johannesburg and Pretoria, South Africa's third major metropolitan area, Durban, followed by some

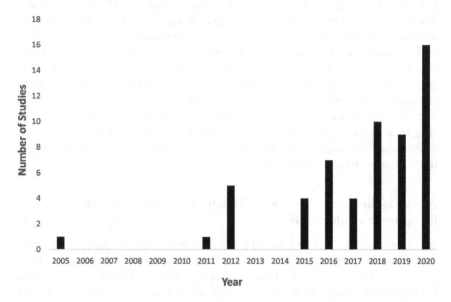

Figure 1.1 Number of studies on tourism and climate change in southern Africa.

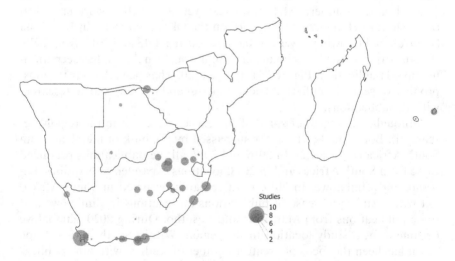

Figure 1.2 Number of studies in different locations across southern Africa.

middle-order cities such as Gqeberha, Richards Bay, and Mbombela, popular beach tourism destinations such as St. Francis Bay and Cape St. Francis (Figure 1.2).

Regarding themes under investigation, perception-based studies of climate change threats are most common, representing 22 publications. Studies on mitigation and sustainability are the second most prolific area of study, with seven publications. Smaller thematic areas include adaptation and policy recommendations (see Hambira, 2018; Hoogendoorn et al., 2021), methodological developments (see Fitchett et al., 2016b), climatic suitability assessments (see Fitchett et al., 2017; Mushawemhuka et al., 2021; Noome & Fitchett, 2019, 2021), and economic analyses (see Fitchett et al., 2016c). Research on tourism and climate change in southern Africa often includes the tourism sector response to sustainable mitigation strategies, greening of the accommodation sector, studies on greenhouse gas emissions from different sub-sectors of the tourism sector, even where these are not the primary focus of the study. A total of six review and theoretical papers have been published for the region thus far, which are usually the most highly cited. These core themes are notably different from the international discourse where themes of vulnerability, snow making, and climatic variability are more common, but there are also some overlapping themes such as indices, sustainability, and perceptions (see Fang et al., 2018).

To explore key terminology in the current discourse, using the titles of the 57 papers in the database, a word cloud has been produced using the software *WordItOut* (Figure 1.3). The most frequently mentioned words in titles of publication are unsurprisingly, 'tourism' and 'climate'. This is followed

Figure 1.3 Word cloud of titles of tourism and climate change publications in southern Africa.

by 'South' and 'Africa', and 'Botswana', 'Southern', and 'African'. Thereafter, a number of words are used with equal frequency including 'operators', 'adaptation', 'impacts', evidence', 'industry', 'policy', 'tourists', and 'nature', Less-frequently used words include 'attitudes', 'plans', and 'events'.

A large proportion of the research on tourism and climate change in southern Africa has been conducted by geographers, but there is considerable diversity in the journals in which the work is published. The most common thematic area of the journals is Tourism, Leisure, and Hospitality, representing 23 publications. Many of these are published in the regional open access journal *African Journal of Hospitality, Tourism and Leisure*, and far fewer in flagship journals such as the *Journal of Sustainable Tourism*, and *Current Issues in Tourism*, and mid-range journals such as *Tourism Review International* and *Journal of Outdoor Recreation and Tourism*. This deviates significantly from international trends, where most papers on tourism and climate change are published in environmental science, environmental studies, climatological, and atmospheric science journals, followed by journals under the theme of Tourism, Leisure, and Hospitality (Fang et al., 2018). Locally, the second largest group of publications (16) is in geographical journals such as the *African Geographical Review, South African Geographical Journal, Bulletin of Geography: Socio-Economic Series* from Poland, and the *Singapore Journal of Tropical Geography*. The prevalence of Geography journals as an outlet for climate change and tourism research is relatively unique to the southern African experience (Fang et al., 2018). A smaller group of papers have been published in development studies journals such as *Development Southern Africa* and regional science journals such as *South African Journal of Science* and *Transactions of the Royal Society of South Africa*. Occasionally, southern African tourism and climate change research has been published in the *International Journal of Biometeorology, Weather, Climate and Society,* and *Environmental Development*.

In terms of the geographical distribution of the authors of research on tourism and climate change in southern Africa, the majority list affiliation with a South African institution (Figure 1.4). The second and third largest representation is from Finland and Botswana, respectively. A smaller contribution comes from authors currently affiliated with institutions in Zimbabwe and Mauritius. This is a notably small geographic representation in authorship, given the broad distribution of countries under investigation and the extensive international discourse on tourism and climate change.

However, the increasing importance of climate change and tourism research within broader southern African tourism and tourism geography research was underlined by Rogerson and Visser (2020) in their recent edited collection *New Directions in South African Tourism Geographies*. Future research should endeavour to involve far greater representation from the rest of southern Africa. Most notably, significant gaps lie in studies on, and authored by residents or nationals of, Mozambique, Namibia, Madagascar,

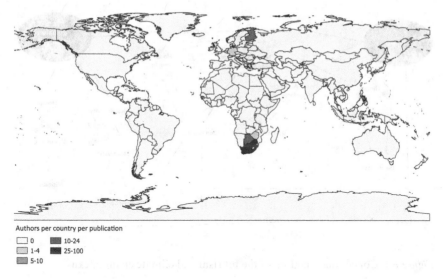

Authors per country per publication

☐ 0   ▨ 10-24
▢ 1-4   ■ 25-100
▨ 5-10

Figure 1.4 Authors per country per publication.

and the smaller Indian Ocean Islands. Contributions from these localities would contribute depth and detail to regional knowledge.

Intersections in tourism and climate change research

The tourism and climate change nexus involves a complex interaction of a variety of variables, spanning the natural and anthropogenic environments (Scott et al., 2012). This can be conceptualised diagrammatically (Figure 1.5), highlighting the highly interdisciplinary nature of tourism and climate change research. Stemming from discrete themes of tourism and climate change are the key elements affected by weather and its components of temperature, sun, rain, wind, snow and humidity, and climatic consequences such as floods, droughts, and storms. Indeed, climate and weather also impact tourism in different ways both in the short and the long term which requires short-term adaptation mechanisms such as the building of retaining walls for flooding, but also the physical moving of infrastructure in the long term (Fitchett et al., 2016a).

These adaptation mechanisms and aspects influence a wide variety of elements such as mitigation, sustainability, and the touristic features of travel, arrivals, and departures, which are ultimately affected by environmental suitability, adaptation, and resilience. These are inter-connected by elements such as economy, infrastructures, and destination image. The

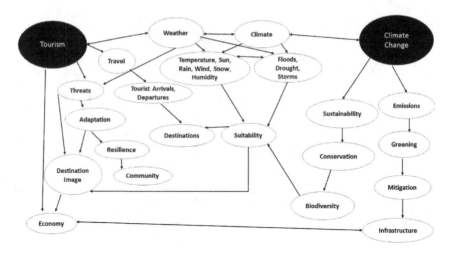

Figure 1.5 Conceptual diagram of the tourism and climate change nexus.

tourism sector also influences the environment quite substantially through direct release of greenhouse gas (GHG) emissions, the overuse of resources, and alteration of natural elements such as diverting of rivers and clearing of land for development.

Structure of this book

This book is structured as follows. After the Introduction, Chapter 2 discusses the climate and climate change of southern Africa, describing the key synoptic features and their inter-relationship in determining the Köppen-Geiger zones. This chapter includes three boxes by local experts, exploring climate change projections, threats of tropical cyclones in the south-west Indian Ocean, and changing rainfall seasonality in the region. In Chapter 3, climate change threats to sustainability will be explored against the backdrop of sustainable tourism development. The chapter involves a box focusing on climate change and community-based tourism. Chapter 4 reflects on the methods used in tourism and climate change research in southern African and methodological advancements originating from the region. In this chapter, the boxes include a more detailed reflection on tourism climate indices, the effective uses of questionnaires, and personal in-depth interviews in the southern African setting.

Chapter 5 considers the risks and impacts of climate change to the southern African tourism sector, with a focus on droughts, flooding, heat, rain, and SLR. The boxes in this chapter focus on the specific threats to coastal tourism, arid tourism, NTB, and cultural tourism. Chapter 6 explores the resilience of the southern African tourism sector, with boxes focusing on policy responses and needs in Botswana and Mauritius. Chapter 7 reflects on the adaption strategies available to southern African tourism and the challenges in implementing effective and timely adaptation. The boxes for this chapter report on adaption in snow tourism, regional infrastructure, and a specific case study of adaptation in Zimbabwe. In Chapter 8, the role of tourism in contributing to climate change is explored, reflecting on GHG emissions through transport and energy use, water use, and the challenges in quantifying and mitigating these impacts. The boxes in the chapter explore the role of the tourism sector in the drought in Cape Town, South Africa, and the significance of transport in contributing to climate change. Chapter 9 considers the importance of climate change and tourism governance and policymaking in relation to SDGs, in the region. The boxes in this chapter will consider SDGs in the tourism and hospitality sector. Chapter 10 synthesises the key findings of the book.

2 Climate and climate change of southern Africa

Introduction

The southern African region offers amongst the world's most favourable climates for tourism, with warm temperatures, frequent sunny skies, and low incidence of prolonged rainfall (Perch-Nielsen et al., 2010). This has contributed to the international marketing of the region as sunny southern Africa (Harms, 2006; Saayman & Saayman, 2008) and has facilitated the development of successful outdoor tourism industries including beach, nature-based, and adventure tourism attractions (Fitchett & Hoogendoorn, 2018). Tourism Climatic Indices, which are used in quantifying the climatic suitability of a destination for tourism, confirm this, with scores ranging from very good through to ideal (Fitchett et al., 2017). The climate of southern Africa is, however, highly spatially and temporally heterogenous, spanning 14 of the Köppen-Geiger climate zones (Beck et al., 2018). A range of extreme climatic events interrupt this 'ideal' climate, with significant and often long-lasting impacts on the tourism sector (Smith & Fitchett, 2020). A detailed understanding of the atmospheric systems that control the climate, the regional climatic zones, and the threats of extreme climatic events are thus of critical importance in understanding the climatic threats to tourism in southern Africa. The projected climate change across this highly heterogenous region is of critical importance in understanding the changing nature of the threats and developing regionally specific resilience through data-informed adaptation.

In broad terms, the climate of southern Africa is influenced by a combination of the latitudinal position, the distance from and properties of the adjacent oceans and the properties of their boundary currents, and the topography and induced altitudinal ranges and orographic effects (Jury et al., 1993; Lennard, 2019). Spanning the tropics to the mid-latitudes, southern African climate is influenced by the tropical easterly wave in the northern regions and the westerly wave in the southern regions and consequently has distinct tropical, sub-tropical, and temperate climatic bands (Desbiolles et al., 2020). This latitudinal position also results in distinct climatic seasonality in both rainfall and temperature regimes (Roffe et al., 2019, 2020; van der Walt & Fitchett, 2020). A broadly semi-arid region, southern Africa

DOI: 10.4324/9781003102618-2

has a distinct east-west moisture gradient across the subcontinent from very moist conditions with up to 1500 mm of rainfall per annum along the eastern boundary of South Africa, Lesotho, Eswatini, and Mozambique through to arid conditions along the west coast of South Africa and much of Namibia (Crétat et al., 2012). This rainfall regime is driven by the difference between the warm Agulhas Current of the southwest Indian Ocean encouraging uplift and supplying moisture to the cold Benguela Current of the southeast Atlantic Ocean inducing subsidence and inhibiting cloud formation (Jury et al., 1993). Rainfall is concentrated along the escarpment of South Africa, inducing orographic cloud formation (Xulu et al., 2020). Temperatures range from very warm tropical to sub-tropical conditions in the low-lying regions of the subcontinent to cold subalpine conditions in the highlands of eastern Lesotho and the South African Drakensberg (Kruger & Shongwe, 2004). This highly complex climate is controlled by a range of processes from local through to global scales (Desbiolles et al., 2020) and influenced by a variety of factors which interact non-linearly (Jury et al., 2004).

In discussing the weather, climate, and projected climate change for a region, it is important to have a clear understanding of the meaning of, and differences between, each of these terms. The weather refers to the atmospheric conditions experienced at a scale of a day or less – the temperature, occurrence and type of precipitation, cloud cover, wind speed and direction, and humidity, which can change throughout the day and are specific to small geographical regions (Lennard, 2019; Zhang, 2013). The climate refers to the average weather that has been experienced over a long period. The World Meteorological Organization (2017) requires a minimum period of 30 years to be considered in determining the regional climatology and climate 'normals'. As succinctly put by Herbertson (1901), and repeated often since (see Lennard, 2019; Zhang, 2013), "climate is what we may expect, weather is what actually we get". The climate can be calculated monthly, seasonally, or at an annual scale (Lennard, 2019). For example, the mean annual temperature on earth during the Holocene (past 11,700 year) is 13°C (Fitchett, 2019); on the day of writing this sentence (11 September 2020), the temperatures in Johannesburg spanned 14°C at night to 26°C in the day with clear skies, while in Cape Town, the skies have been partly cloudy with temperatures of 14°C at night and 22°C in the day. The average for neither location is 13°C, but in mid-winter it is often well below it. While the weather of a location fluctuates throughout the day, the climate likewise fluctuates inter-annually (Zhang, 2013). Shorter-term fluctuations over a period of a few years to a few decades, which often cycle around the long-term mean, are referred to as climate variability, while more unidirectional and longer-term changes over decades to centuries are referred to as climate change (Folland et al., 2002).

Climatic zones of southern Africa

Regional climatic conditions can be classified into discrete zones in which temperature, rainfall, and other climatic variables fall within specific ranges or exceed particular thresholds (Engelbrecht & Engelbrecht, 2016). These are useful in determining patterns of relative homogeneity within these highly complex natural systems, which allows for climatic suitability, challenges, and adaptation responses determined for a location to be up-scaled (Beck et al., 2018; Rohli et al., 2015). Regional climatic classifications are often conducted ad hoc for a particular application of interest, and while useful to that application, arguably lead to confusion rather than consensus (Roffe et al., 2019). The Köppen-Geiger classification of climate zones, developed originally by Wladimir Köppen (1936) and updated by in 1961 by Rudolf Geiger, remains the most extensively used regional climate classification system (Engelbrecht & Engelbrecht, 2016; Kottek et al., 2006). The Köppen-Geiger classification system considers long-term monthly, seasonal, and annual temperature and rainfall and broadly represent the biomes that the threshold climates support (Engelbrecht & Engelbrecht, 2016; Rohli et al., 2015). Over the past two decades, following the development of global climate datasets, there has been renewed effort to re-map the contemporary Köppen-Geiger zones and to project future distributions of these zones under climate change (see Beck et al., 2005, 2018; Engelbrecht & Engelbrecht, 2016; Rohli et al., 2015; Rubel & Kottek, 2010).

The Köppen-Geiger classification includes five high-level climate types: tropical, arid, temperate, snow and boreal, and polar, each of which are ascribe a letter code of A to E, respectively (Beck et al., 2005; Rubel & Kottek, 2010). Each of these are then divided into subclasses which reflect the seasonal characteristics, which are coded by further letter codes adjoining the class code (Beck et al., 2005). Only climate types A–C are observed in southern Africa and of the 16 subtypes of these broad groups, 14 are recorded over this region (Figure 2.1), calculated for the period 1980–2016, and 11 South of 22°S (Engelbrecht & Engelbrecht, 2016, calculated for the period 1961–1990). Part of the Köppen-Geiger classification relates to the season of precipitation. Located in the Southern Hemisphere, southern Africa experiences the austral seasons, with summer approximately spanning December through March and winter June through August (van der Walt & Fitchett, 2020). There is, however, considerable heterogeneity in these seasonal distinctions, whether classified by temperature (van der Walt & Fitchett, 2020) or rainfall (Roffe et al., 2020, 2021).

The tropical climate types are found almost exclusively in Mozambique, Madagascar, and the Comoros, with the exception of a very small area of northern Zimbabwe and of the KwaZulu-Natal Province of South Africa (Figure 2.1). This climate zone is largely influenced by sub-tropical to tropical latitudes, and the warm Agulhas Current in the adjacent Indian Ocean. Tropical savanna (Aw) is the dominant subtype, with a region of tropical rainforest (Af) along the east coast of Madagascar and across Mauritius and small pockets of tropical monsoon (Am) in Mozambique and Madagascar.

The arid climate zone covers the largest area of southern Africa, extending from the west coast of South Africa and Namibia and covering all but a

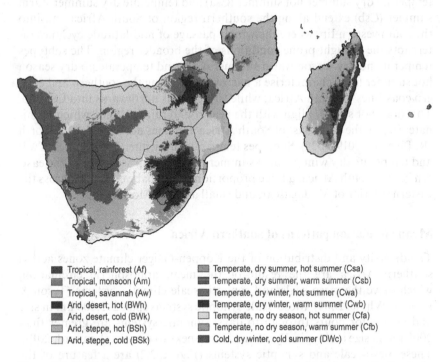

Tropical, rainforest (Af)
Tropical, monsoon (Am)
Tropical, savannah (Aw)
Arid, desert, hot (BWh)
Arid, desert, cold (BWk)
Arid, steppe, hot (BSh)
Arid, steppe, cold (BSk)

Temperate, dry summer, hot summer (Csa)
Temperate, dry summer, warm summer (Csb)
Temperate, dry winter, hot summer (Cwa)
Temperate, dry winter, warm summer (Cwb)
Temperate, no dry season, hot summer (Cfa)
Temperate, no dry season, warm summer (Cfb)
Cold, dry winter, cold summer (DWc)

Figure 2.1 Köppen-Geiger zones of southern Africa (adapted from Beck et al., 2018).

small isolated region of Namibia, all of Botswana, more than half of South Africa, and regions of Zimbabwe, Mozambique, and Madagascar (Figure 2.1). The subtype arid steppe hot (BSh) covers the greatest land surface area into each of the aforementioned countries, spanning the central and northern regions of the subcontinent and the southwestern coast of Madagascar. Subtypes arid desert hot (BWh) and arid desert cold (BWk) form the majority of the western coastline of the subcontinent and extend towards the central interior spanning the southern half of Namibia, the southwestern quadrant, and central interior of Botswana, and the western third of South Africa. While this aligns with the position of the Namib and Kalahari Deserts and the Karoo and is driven largely by the subsidence-inducing cold Benguela Current in the adjacent Atlantic Ocean, the region is larger than those alone. A small area of this subtype is also calculated for the confluence of Zimbabwe, Botswana, and South Africa, which is notably not a desert but rather sits along the Limpopo River. A relatively wide band of subtype arid steppe cold (BSk) is found in South Africa west of Lesotho, bordering the zones of BWh and BWk.

The temperate climate type (C) largely fills the eastern-central regions between those classified as tropical (A) and those as arid (B). The subtypes temperate dry summer hot summer (Csa) and temperate dry summer warm summer (Csb) extend along the southern region of South Africa and into the southwestern tip, where the winter passage of mid-latitude cyclones interrupts the drought-prone conditions of the broader region. The subtypes temperate no dry season warm summer (Cfb) and temperate no dry season hot summer (Cfa) characterise a narrow band along the southern and eastern coastlines of South Africa, while Cfb extends northwards into Lesotho. This does not strictly align with the year-round rainfall zone, which terminates before the east coast of South Africa and does not extend as far north (Roffe et al., 2019, 2020). Subtypes temperate dry winter hot summer (Cwa) and temperate dry winter warm summer (CWb) comprise much of the eastern half of South Africa, a large proportion of Zimbabwe, a band across the eastern interior of Madagascar, and small areas of Mozambique.

Mean circulation patterns of southern Africa

The diversity and distribution of the Köppen-Geiger climate zones across southern Africa are the result of the mean atmospheric circulation, which drive the meso-scale and synoptic-scale climatic systems (Tyson & Preston-Whyte, 2004). Meso-scale synoptic systems span 2–2,000 km in size and persist for hours through to days; synoptic systems are greater than 2000 km in size and typically persist from weeks to months (Lennard, 2019). These meso-scale and synoptic systems (Figure 2.2) are a feature of the broad-scale interaction between the atmosphere, hydrosphere (incorporating the water bodies, particularly the oceans), and the lithosphere (incorporating the earth's surface, Xulu et al., 2020).

A semi-permanent zone of convergence, and uplift, characterises the meeting of the northern and southern Hadley cells, at a region termed the Inter-Tropical Convergence Zone (ITCZ). There is considerable seasonal variation in the location of this zone and the resultant band of rainfall, as this marks the location of maximum solar radiation (Manatsa & Reason, 2017). There is also more local spatial variation in the position, driven by orography and land-ocean temperature variations (Jury et al., 1993). The shifting position of the ITCZ influences the extent and timing of intra-annual shifts in each of the other synoptic features, which induces the distinct seasonality (Cook et al., 2004). The ITCZ is associated with strong uplift, which is responsible for the majority of the rainfall along the northern regions of southern Africa. The Congo Air Boundary (CAB) extends broadly west from the southern most regions of the ITCZ over the subcontinent towards the northern-most regions of the Namibian coastline (Tyson & Preston-Whyte, 2004). This region of convergence marks the meeting of the warm moist Indian Ocean air transported via the tropical easterlies and the cool dense Atlantic Ocean air from the transported by westerlies that form from recurved Atlantic easterlies (Howard & Washington, 2019). The meeting of these two distinct air masses forms a very deep zone of convergence

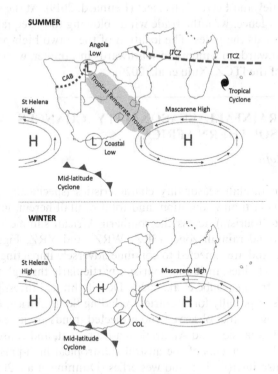

Figure 2.2 Synoptic climate systems of southern Africa.

and uplift. The CAB is most clearly defined in the austral spring and early summer, and the exact location varies on a daily basis, representing the region of preferential formation of convective low pressure cells and inducing heightened moisture through resultant rainfall (Howard & Washington, 2019; Tyson & Preston-Whyte, 2004).

Among the two largest systems in terms of their spatial influence are the semi-stationary St Helena High Pressure Cell situated over the south Atlantic and the Mascarene High over the South Indian Ocean (Desbiolles et al., 2020). These high-pressure cells are characterised by the subsidence of air, largely inhibiting cloud formation, and as a consequence, rainfall (Xulu et al., 2020). As a function of Coriolis force, these systems are characterised by anti-clockwise air circulation. This results in the movement of warm moist air off the Indian Ocean over the eastern region of the southern African subcontinent inducing rainfall, and the movement of air away from the west coast, inducing the upwelling of cold water along the southeastern boundary of the Atlantic Ocean, reducing temperatures of the Benguela Current (Abba Omar & Abiodun, 2020). These two high-pressure systems form part of the sub-tropical subsidence zone which runs across the

Southern Hemisphere, which develops due to the upper air convergence of air as the Hadley and Ferrel Cells meet (Lennard, 2019). At the surface, this results in divergence, with the trade winds blowing to the equator, and the westerlies towards the poles. The location of these two high-pressure cells migrates northwards in winter and southwards in summer, with the shifts in the position of the ITCZ (Xulu et al., 2020).

BOX 2.1 RAINFALL SEASONALITY CHANGES ACROSS SOUTHERN AFRICA

Sarah J. Roffe

Changes in rainfall seasonality characteristics (specifically considering wet-season timing, duration, and total rainfall herein, given their relevance to tourism) across the southern African summer-, winter-, and year-round rainfall zones (SRZ, WRZ, and YRZ; Figure 2.1.1) have begun and are expected to continue adversely impacting rainfall-dependent activities, including tourism, particularly through water resource constraints (Dube et al., 2020). Despite limited investigation of such trends, especially for countries excluding South Africa, emerging evidence suggests that observed and projected trends are in response to Hadley cell expansion and Antarctic sea-ice retreat, and consequential poleward displacements of the austral sub-tropical high-pressure belt (which is also intensifying) and westerlies (Dunning et al., 2018; Roffe et al., 2021). Quantified trends are typically insignificant, and magnitudes are inconsistent. Therefore, only reported trend directions are considered, however, as these are occasionally inconsistent, a broad summary considering patterns of consistent trend directions per rainfall zone across observed and projected trend investigations follows.

Across the SRZ, spanning most of southern Africa and where convective rainfall occurs primarily during September/October–March/April, many authors report trends of a reduction in wet-season (or for months therein) rainfall totals with an increase in dry-spell periods (Roffe et al., 2021). Despite this, slight increases in wet-season (or for months therein) totals are consistently detected for western Madagascar and Mauritius (Dunning et al. 2018; Pascale et al., 2016; Senapathi et al., 2010). Trends for the wet-season timing and length are mostly consistent, where these primarily indicate a tendency towards shorter wet-seasons largely due to later start-dates and earlier end-dates (Roffe et al., 2021).

The most consistent trend for the April–September WRZ wet-season is for reduced frontal rainfall totals, which are expected to persist with an increased frequency of prolonged dry-spells. Despite weak consistency between observed and projected wet-season start- and end-date and duration trend directions (Dunning et al., 2018; Roffe et al., 2021), reductions in autumn and spring rainfall are detected, tentatively suggesting a shorter wet-season duration due to later start-dates and earlier end-dates, respectively. Notably, these seasonal trends correspond to wet-season projections (Dunning et al., 2018; Pascale et al., 2016).

Between the SRZ and WRZ, where the YRZ (i.e. year-round rainfall occurs from summer and winter rain-bearing systems) occurs, distinct zones exist for wet-season trends. Across the southern coast, trends reflect later start-dates and earlier end-dates, resulting in a reduced wet-season duration. While observed trends reflect a slight increase in wet-season (specifically winter months therein) totals (Roffe et al., 2021), spring and summer declines are prominent and contribute strongly to the shorter wet-season trend (Mahlalela et al., 2020). Interior YRZ trends primarily indicate later start- and end-dates, resulting in a shorter wet-season duration, and these coincide primarily with trends towards reduced wet-season totals.

The consistent wet-season trends for each rainfall zone (Figure 2.1.1) present a concerning future for rainfall-dependent activities. For tourism specifically, reductions in wet-season totals and increases in dry-spell periods, which are detected for most of southern Africa, are, and will continue to be, detrimental to economic stability for this industry; as has been observed for Cape Town, South Africa, during the 2015–2017 'Day Zero' drought (Dube et al., 2020).

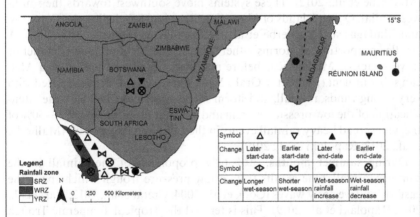

Figure 2.1.1 Map of southern African (Africa south of 15°S) summer-, winter-, and year-round-rainfall zones (boundaries adapted from Roffe et al., 2019) and consistent wet-season trends per rainfall zone compiled based on literature presented herein. For the SRZ specifically, wet-season timing and duration trends are presented on mainland southern Africa, while deviations for wet-season totals are demonstrated where necessary.

More localised and often shorter-lived low pressure cells are found across southern Africa. Some of these are warm-cored and result from local heating, whereas others are cold-core systems which form in the mid-latitudes and migrate northwards (Sousa et al., 2018). Both types are associated with rainfall. The Botswana Low is a warm-cored semi-stationary tropical low situated just north of Botswana (Tyson & Preston-Whyte, 2004). Mid-latitude cyclones form at the polar front and migrate through the Ferrel cell year-round, driven by the temperate westerlies (Abba Omar & Abiodun, 2020). As the ITCZ moves northwards during the austral winter, these systems can extend further equator-wards, with a greater proportion passing over the southern tip of the subcontinent (Roffe et al., 2019). These systems are responsible for the winter rainfall experienced in the southwestern Cape (Burls et al., 2019; Sousa et al., 2018). Some of these systems which form from unsteady Rossby waves form troughs which extend equator-wards and are interrupted by the high-pressure cells (Xulu et al., 2020), resulting in a portion which is detached or cut off from the westerlies – these are termed cut-off lows (COLs) and are usually responsible for more extreme rainfall, with the potential for flooding (Abba Omar & Abiodun, 2020; Lennard, 2019) and snow in the interior of South Africa during winter (Stander et al., 2016). Short-lived coastal lows form along the South African coastline preceding mid-latitude cyclones where they are trapped by the escarpment, driving the berg wind conditions which are often responsible for fires in the region (Lennard, 2019). Tropical Depressions form over the southwest Indian Ocean during the summer months and, under a set of conducive ocean-atmospheric conditions, intensify to tropical storms and tropical cyclones (Malherbe et al., 2012). These systems move southwest towards the southwest Indian Ocean Islands of Mauritius, Reunion, Seychelles, the Comoros, and Madagascar (Malherbe et al., 2012; Pillay & Fitchett, 2019, 2021). A smaller proportion of storms either form in the Mozambique Channel or traverse across Madagascar before making landfall on the southern African subcontinent (Fitchett & Grab, 2014). These systems are associated with very strong winds, rainfall, and storm surges, which escalate with the intensification of the low pressure system and account for estimates of 28–50% of the widespread heavy rainfall events in the Limpopo River basin (Malherbe et al., 2012; Rapolaki et al., 2019).

During summer months, 30%–60% proportion of the rainfall in the summer-rainfall zone results from a low pressure trough which spans the easterly and westerly waves (Cook et al., 2004; Crétat et al., 2012; Hart et al., 2013; Rapolaki et al., 2019). This is termed the Tropical Temperate Trough (TTT), which is characterised by an intermittent NW–SE cloud band stretching from north-eastern Namibia through to the south coast of South Africa (Rapolaki et al., 2019). The strength and frequency of development of the TTT cloud bands are influenced by sea surface temperatures in both the Indian and Atlantic Oceans and the land surface temperature of the Kalahari (Manatsa & Reason, 2017). Variability in the strength and timing of the Mozambique Channel trough, a convective region in the central and southern Mozambique Channel, further influences the location and strength of the TTTs (Barimalala et al., 2020). The tropical-sub-tropical South Indian Convergence Zone, which forms the preferred region of formation of the TTT (Hart et al., 2018), is unique in the Southern Hemisphere systems in that the position and topography of Madagascar prevents direct moisture intrusion into the system from the southwest Indian Ocean, which rather has to first move around Madagascar (Barimalala et al., 2018; James et al., 2020). Over the Atlantic Ocean, the position of the Angola-Benguela Frontal Zone and the strength of the Angolan Low have an impact on TTTs (Desbiolles et al., 2018, 2020). The Angolan Low itself is responsible for nearly 15% of the rainfall in Namibia (Desbiolles et al., 2020), while warm sea surface temperature events in the south Atlantic off Angola are responsible for heightened rainfall and the potential for flooding in Namibia during the months of February to April (Reason & Smart, 2015). These changes in the strength and position of the TTT influence the rainfall of Namibia, Botswana, Zimbabwe, Mozambique, South Africa, and Lesotho (Desbiolles et al., 2020). There is also a low incidence of TTTs forming in association with COLs, recorded in 28 of 302 continental TTTs over the period 1979–1999 (Hart et al., 2013).

Extreme climatic events in southern Africa

In addition to these synoptic-scale features which affect the intra-annual seasonal climates of southern Africa, the climate of the region is also influenced by large-scale forcing mechanisms driven by fluctuations in major ocean-atmospheric circulation systems and their feedbacks, which drive extreme climatic conditions. The largest-scale forcing mechanism is the El Niño Southern Oscillation (ENSO), characterised by a 7–14-year cycle in temperatures of the equatorial Pacific. In broad terms, El Niño events result in drier conditions along much of the summer rainfall region of the subcontinent but drought conditions in the winter-rainfall zone; during La Niña events, the converse occurs (Crétat et al., 2012; Malherbe et al., 2020). This is in part due to the impacts on the South Indian Convergence Zone through the weakening of the Mascarene High through a wave response forced through the east Pacific (Hart et al., 2018). During La Niña events, 150%–200% more TTT cloud bands develop over the subcontinent, contributing to the heightened rainfall in the summer-rainfall zone during La Niña, while the converse occurs during El Niño contributing to the drought conditions (Hart et al., 2018). At a more local scale, the climate is influenced by the South Indian Ocean Dipole, the Southern Annular Mode, and the Benguela Niño. Upper air features including the Quasi-Biennial Oscillation, the Madden–Julian Oscillation, and local features including the 18-year Dyer–Tyson rainfall cyclicity further influence both the mean climatic conditions and the incidence of extreme climatic events (Crétat et al., 2012; Hart et al., 2013).

The largest scale extreme climatic events experienced in southern Africa are droughts and floods. Of these, the former persist for months to years, while the latter are short-lived events of at most a few days, but have impacts which are felt over similar periods to droughts. One of the primary causes of recurrent droughts and flooding in the region are regional displacements of moisture during El Niño and La Niña (Crétat et al., 2012; Malherbe et al., 2020) and positive and negative Indian Ocean Dipole (IOD) events (Jury et al., 2004). Under climate change, progressive expansion of the Hadley cell and resultant changes in the distribution of moisture corridors is heightening drought effects in the winter-rainfall zone, as a result of the poleward displacement of the westerlies and the consequent reduction in mid-latitude cyclone passage (Burls et al., 2019; Sousa et al., 2018). This has been posited as the primary cause of the 2015–2017 'Day Zero' drought in Cape Town, South Africa, during which rainfall dropped to <50% of the long-term normal, and dams had dropped to 21% capacity marking it as the worst water shortage in the region in 113 years (Abba Omar & Abiodun 2020; Burls et al., 2019). While there is some disagreement as to the significance of the role of water use management in this drought relative to climate change (Muller, 2018; Otto et al., 2018), the climatic anomalies have been well documented (Burls et al., 2019; Sousa et al., 2018).

Through similar mechanisms, there has been a southward expansion in the region of tropical cyclone storm tracks, which heightens the incidence of tropical storm and cyclone landfall and influence in southern Africa, resulting in flooding across northern regions of South Africa, Mozambique, Botswana, and Zimbabwe (Fitchett, 2018; Pillay & Fitchett, 2019). ENSO and IOD also influence the distribution and frequency of tropical cyclones (Fitchett & Grab, 2014). Another source of flooding is COLs, as described earlier in this chapter. While 46% of COLs are associated with any rainfall at all, 22% of COLs are associated with extreme rainfall and comprise a significant source of extreme rainfall events (Abba Omar & Abiodun, 2020) and widespread snowfalls in southern Africa (Stander et al., 2016). COLs occur most frequently around the southwestern Cape and between the months of June through August (Abba Omar & Abiodun, 2020). A COL was responsible for the Laingsburg flood of 1981 (Singleton & Reason, 2007), the KwaZulu-Natal flood of 1987, and the Montagu flood of 2003 (Lennard, 2019), while tropical cyclones Idai and Eline have been responsible for two of the greatest humanitarian disasters in Mozambique (Emerton et al., 2020).

BOX 2.2 TROPICAL CYCLONE THREATS TO SOUTHERN AFRICA AND THE SURROUNDING ISLANDS

Micheal T. Pillay

Tropical cyclones are large, intense rotating storm systems with wind speeds >119 km/hr identifiable by an eye, surrounded by thunderstorm cloud bands with a diameter ranging from 200 km to 1,000 km. Tropical cyclones are accompanied by devastating winds, extreme rainfall, storm surges, high waves, and occasionally flooding in coastal areas (Gori et al., 2020). With the highest intensity systems having wind-speeds exceeding 300 km/hr, accompanied by heavy rainfall, these systems are associated with many hazards and the capacity to cause widespread damage (Altman et al., 2018). Tropical cyclones are among the costliest natural disasters in the world, accounting for a third of all natural disaster-related damages. Approximately, ZAR14.5 trillion in damage was caused by tropical cyclones globally from 2000 to 2017 (Ye et al., 2020). Loss of life, farmland, infrastructure, and livelihoods also result, which disproportionately impacts people living in vulnerable and socio-economically disadvantaged countries, including southern Africa (Peduzzi et al., 2012). Tourism sectors of island nations are particularly vulnerable, as these socio-economically challenged regions cannot rebuild or recover easily, and tourist activity could be discouraged if and when such hazards increase (Khazai et al., 2018).

Tropical cyclones form and intensify into mature system under specific environmental conditions. These include sea surface temperature above 26.5°C–27°C to a depth of 50 m, atmospheric instability, which promotes strong surface convergence and rising of air masses, high relative humidity which acts as a source of heat and moisture for the storm, and low wind shear which promotes the vertical formation of a tropical cyclone structure (Klotzbach et al., 2017). Usually, these conditions are found in tropical regions of the ocean from 5° latitude, where the effect of Coriolis force is sufficient to allow air mass rotation (Liang & Chan, 2005). These tropical cyclones mature over the ocean and usually only dissipate when they make landfall or encounter cooler ocean regions. They are steered by global-scale winds and sea surface temperature isotherms (Pillay & Fitchett, 2019). Coastal regions are therefore impacted the most frequently by landfall, but due to the size of the systems, even areas far inland can be affected by strong winds and rainfall (Mendelsohn et al., 2012).

Climate change is drastically affecting ocean basins and their tropical cyclone climatology (Altman et al., 2018). Tropical cyclone intensity is increasing, and while overall frequency has decreased, the number of high-intensity tropical cyclones has increased (Song et al., 2018). This is also true in the southwest Indian Ocean, where category 5 tropical cyclones (highest intensity based on the Saffir Simpson scale) have recently emerged and are occurring increasingly frequently (Fitchett, 2018). The locations of high-intensity tropical cyclones are shifting polewards (Figure 2.2.1) as the high sea surface temperature environment which support them is becoming more common further south of the tropical regions (Altman et al., 2018; Pillay & Fitchett, 2019). This heightens the probability of high-intensity storms influencing the southern African mainland, which has previously been protected from such systems by Madagascar's buffering effect (Pillay & Fitchett, 2019). Globally, it is predicted that R290 Billion in damages can be expected annually by the year 2100 (Narita et al., 2009). Countries with low adaptive capacity, such as those in Africa and surrounding islands, could face much more risk and damages to infrastructure, economy, and livelihoods in the future (Peduzzi et al., 2020).

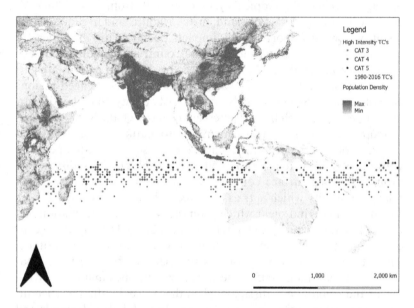

Figure 2.2.1 Category 3 to 5 tropical cyclone distribution over the last 30 years in comparison to the 2010–2016 season (all categories). The continental shading indicates population density ranging from high to low.

Extreme Temperature Events (ETEs) are of shorter duration and often smaller geographical scale than droughts and floods, but have a significant impact on human health and comfort (van der Walt & Fitchett, 2021a, 2021b). ETEs can be defined by either the exceedance of human comfort level threshold conditions or statistically anomalous temperatures for a given location and season or a combination of the two (van der Walt & Fitchett, 2021c). They can also be calculated at a micro-climatic scale such as the effects of extreme temperatures at a bus stop or school playground (Vanos et al., 2016) or at the scale of a city represented by a meteorological station or a grid cell (Mackellar et al., 2014). When considering thermal comfort and stress, temperature is considered together with the relative humidity, whereas statistical measures of anomalous temperature often evaluate temperature alone (van der Walt & Fitchett, 2021c). Both cold and hot ETEs are currently recorded in southern Africa (Kruger & Sekele, 2013; van der Walt & Fitchett, 2021a, 2021b). The emission of pollutants, including GHGs, serves to induce a localised influence of heat trapping over cities, while artificial surfaces alter the albedo and heating of buildings directly contributes heat emission (Fitchett et al., 2020; Goldreich, 2006). This is heightened in environments of high pollutant load, and of stable atmospheric conditions, thus inducing warmer microclimate conditions over the Gauteng Province of South Africa and its surrounds (Fitchett et al., 2020), reducing the incidence of cold ETEs and intensifying warm ETEs.

Less common extreme climatic events in southern Africa include (1) annual snowfalls, which are largely constrained to the high-altitude regions of the Drakensberg Maloti spanning South Africa and Lesotho and the Cape Fold Mountains in the southwestern Cape of South Africa (Stander et al., 2016), (2) hail storms particularly in the interior plateau of southern Africa, and (3) tornadoes predominantly across the Gauteng, KwaZulu-Natal, Mpumalanga, Free State and Eastern Cape provinces of South Africa, with low frequency of occurrence in Lesotho and Eswatini (Goliger & Retief, 2007). Snowfalls in southern Africa are seldom sufficiently severe to be classified as blizzards, which would formally constitute extreme climatic events (Stander et al., 2016), and instead serve as an attraction for tourists within southern Africa (Hoogendoorn et al., 2021). Hail storms are common in the central plateau of southern Africa, driven by strong convection in summer months (Pienaar et al., 2015). Occasionally, severe hail storms occur, resulting in damage to vehicles and properties and resultant insurance claims (Pienaar et al., 2015). Tornadoes are very infrequent by global standards and relatively low in severity, but an important consideration for the insurance of building and infrastructure for tourism destinations (Goliger & Retief, 2007; Milford et al., 1994).

Climate change projections for southern Africa

In the 21st century, historical climate change can be calculated over past decades from meteorological data, while continued climate change can be modelled into the future through statistical or dynamic climate models downscaled to the region of interest (Mackellar et al., 2014). For much of the African continent, studies of historical climate change over the past century are hampered by the poor availability of high spatial resolution, accurate data (Mackellar et al., 2014). South Africa has one of the most robust climate data networks on the continent (Ziervogel et al., 2014), yet many stations have less than 30 years of continuous data, preventing the calculation of climate normals and climate change trends (van der Walt & Fitchett, 2021c).

BOX 2.3 PROJECTED CLIMATE CHANGE FUTURES OF SOUTHERN AFRICA

Francois A. Engelbrecht

Southern Africa (here defined as Africa south of 15°S) is located in the subtropics and consequently has a warm climate that displays pronounced dry-wet seasonality. The far western parts of southern Africa are arid, with semi-arid regions stretching eastwards into the central interior (Engelbrecht & Engelbrecht, 2016). Summer rainfall in the region is highly variable and strongly influenced by the ENSO. Severe droughts typically occur in association with the El Niño events, whilst years with above normal rainfall and in some cases flooding are usually associated with La Niña events (Archer et al., 2018). The southwestern Cape and Cape south coast regions of South Africa are winter and all-year rainfall regions, respectively (Engelbrecht et al., 2015a), where periods of low winter rainfall tend to co-occur with the positive phase of the Southern Annular Mode (SAM).

Temperatures in southern Africa have been rising drastically over the last six decades at a rate of about twice the global rate of temperature increase (Engelbrecht et al., 2015b). The frequency of occurrence of ETEs such as heat-wave days and high fire-danger days has similarly increased substantially (Kruger & Sekele, 2013). Statistically significant trends in rainfall have only been identified over relatively small areas including decreasing trends over the southern coastal and eastern interior regions of South Africa (Kruger, 2006). Increasing rainfall trends have been detected over parts of the western Karoo of South Africa (Kruger, 2006).

Under low mitigation futures, drastic warming is projected to continue to occur across the southern African region, reaching values of 4°C–8°C in annual average temperature increases by the end of the century relative to pre-industrial climate (Lee et al., 2021). The region is also likely to become generally drier (Hoegh-Guldberg et al., 2018), with the amplitude of drying strengthening over time (Figure 2.3.1). These changes are the consequence of the strengthening of the sub-tropical high-pressure belt over southern Africa, including the more frequent occurrence of mid-level highs and subsidence in summer (Engelbrecht et al., 2009). Convective cloud formation is suppressed in summer due to the enhanced subsidence, whilst frontal systems are displaced to the south in winter. The circulation changes also imply the more frequent occurrence of heat-waves and an increased likelihood of the occurrence of multi-year droughts. Changes in the amplitude and frequency of El Niño and La Niña events are uncertain in a warmer world, but southern Africa is projected to become generally drier irrespective of changes in ENSO (Lee et al., 2021).

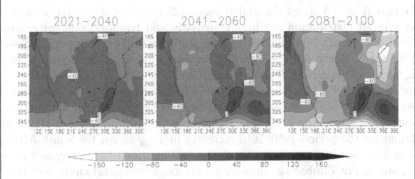

Figure 2.3.1 Projected change in annual average rainfall totals (mm, ensemble average) over southern Africa for the periods 2021–2040 (left), 2041–2060 (middle), and 2081–2100 (right) relative to 1850–1900 (approximating pre-industrial temperature) from an ensemble of 30 global climate models (GCMs) of the Coupled Model Intercomparison Project Phase Six (CMIP6) under a the low mitigation Shared Socio-economic Pathway 5-8.5 (SSP5-8.5).

The Intergovernmental Panel on Climate Change (IPCC) in its Special Report on Global Warming of 1.5°C (Hoegh-Guldberg et al., 2018) consequently identified southern Africa as a climate change hot spot, given that options for adaptation are limited when a water-stressed region becomes even hotter and drier. In fact, should global warming reach a value of 3°C, a regional tipping point may be reached in southern Africa, where both the maize crop and cattle industry in the interior may collapse (Hoegh-Guldberg et al., 2018). Risks of 'Day-Zero'-type multi-year droughts will also increase as regional warming strengthens not only for the southern coastal cities of Cape Town and Gqeberha in South Africa (Otto et al., 2018) but also in the summer-rainfall region.

Despite the general drying in southern Africa in a warmer world, intense thunderstorms may occur more frequently in the warmer climate of the summer-rainfall region (Seneviratne et al., 2021). The projected changes in the tracks and frequencies of landfalling tropical lows and cyclones over Mozambique and northeastern South Africa are uncertain. However, in a warmer world, these systems are likely to hold more water vapour and cause more rainfall than in the past, whilst the more frequent occurrence of intense tropical cyclones is plausible (Hoegh-Guldberg et al., 2018).

Over the past five decades, temperatures across southern Africa have increased by 1.5 times the global mean (Ziervogel et al., 2014). Changes in rainfall distribution are highly spatially and temporally heterogenous (Mackellar et al., 2014). Exploring temperature trends over the period 1960–2003 for 26 South African Weather Services stations across South Africa, Kruger and Shongwe (2004) found statistically significant positive trends in mean annual T_{max} for 13 stations and in mean annual T_{min} for 18 stations. At a seasonal scale, the greatest increases were calculated for autumn and the smallest for spring (Kruger & Shongwe, 2004). Mackellar et al. (2014) found for the period 1960–2010 increases in T_{max} across the South Africa and increases in T_{min} except for the central interior where decreases were recorded. For Zimbabwe, an older yet far longer temporal analysis of temperature trends has been conducted spanning 1987–1993 (Unganai, 1997). A mean increase in T_{max} is reported, centred in two warming periods: the first from the mid-1930s to late-1940s and the second commencing in the 1980s (Unganai, 1997). An update for southern Zimbabwe which extends the records to 2015 reports that T_{max} for the reference period 2000–2015 is 1.7°C warmer than the period 1897–1904 and T_{min} 2.1°C warmer (Sibanda et al., 2018). Similar increases in T_{min} at faster rates than T_{max} are recorded for southern Botswana for the period 1961–2010, while in the northern regions, the T_{max} rates far exceeded those of T_{min} (Mphale et al., 2018). The greatest increases in temperatures in Botswana are recorded for the winter season (Byakatonda et al., 2018a, 2018b). There is a paucity of analyses of historical temperature and rainfall records for Mozambique and Namibia, with the exception of a low spatial resolution analysis of temperature trends for southern and western Africa (New et al., 2006).

Using the WMO/CLIVAR Expert Team on Climate Change Detection and Indices (ETCCDI), Kruger and Sekele (2013) provided the initial assessment on trends in extreme temperatures in South Africa, evaluating data from 28 South African Weather Service Stations for the period 1962–2009. Overall, the incidence of warm ETEs was found to have increased, while cold extreme events decreased, and more severe trends were found for the western, northeastern, and eastern regions of the country (Kruger & Sekele, 2013). These individual ETEs were found to not always be associated with long-term climatic trends (Kruger & Sekele, 2013). A more recent analysis of ETEs demonstrates increases in the incidence of extreme heat events across the county over recent decades and considerable heterogeneity in trends for extreme cold events (van der Walt & Fitchett, 2021a, 2021b). Research on extreme temperature event trends for the broader subcontinent remains under-researched (van der Walt & Fitchett, 2021c).

Climate projections for the region of southern Africa south of 22°S have been used in determining likely changes in the regional climate regimes as classified by the Köppen-Geiger zones (Engelbrecht & Engelbrecht, 2016). Using the A2 scenario of the IPCC SRES under 1°C, 2°C, and 3°C warming, an expansion of the arid hot desert (BWh) and arid steppe hot (BSh) subtypes is projected, while the arid cold desert (BWk) and arid steppe cold (BSk) zones contract. These projected changes would favour the intrusion of trees into grassland regions, which would progressively convert to savannas, while the temperate dry summer regions are projected to contract, replaced by the arid hot steppe zone which would heighten the probability of fires (Engelbrecht & Engelbrecht, 2016). These changes are consistent with much coarser-resolution global-scale projections of changes in the Köppen-Geiger zones under climate change (Beck et al., 2018). This poses significant threats to some of the remaining pristine grasslands of southern Africa and the fynbos region of high endemism (Engelbrecht & Engelbrecht, 2016).

The incidence of droughts and floods is projected to increase under climate change (Ziervogel et al., 2014). Drought events, driven in southern Africa by shifts in the moisture corridors and the role of ENSO, are projected to occur more frequently and intensify in severity throughout the 21st century (Malherbe et al., 2020; Otto et al., 2018). In particular, 'Day Zero'-type droughts in the southwestern Cape are projected to reoccur as the poleward migration of the westerlies continues (Otto et al., 2018; Sousa et al., 2018). The effects of these meteorological droughts will continue to be exacerbated by increased urbanisation and associated anthropogenic water consumption (Otto et al., 2018). Tropical cyclones, which represent one of the most severe sources of flooding, are projected to increase in intensity and are expanding polewards in their range, but have not increased in frequency (Fitchett, 2018; Fitchett & Grab, 2014; IPCC, 2021). The changing dynamics heighten the risk of southern African countries to tropical cyclone landfall and the effects of the storm (IPCC, 2021). Changes in the strength of the Angola low, induced by Atlantic Ocean sea surface temperatures, may have significant impacts on the TTTs and the associated rainfall (Reason & Smart, 2015), while changes in the strength of the Mascarene High will influence the formation of COLs (Xulu et al., 2020). SLR, resulting from the melting of ice sheets and glaciers and the expansion of the water due to warming, heightens the risk of coastal flooding through gradual inundation of the shoreline and through more severe storm surges (IPCC, 2021; Mather & Stretch, 2012).

Of critical importance in modelling and communicating future climates is the acknowledgement and quantification of the uncertainty in the model output (Ziervogel et al., 2014). Although there have been considerable efforts to improve the capacity for climate modelling in southern Africa, and model intercomparisons form a key feature of the IPCC reports (see IPCC, 2021), James et al. (2020) argue that there remains insufficient research into the fidelity of downscaled global climate models when simulating African regional climates. These result in differences between modelled and observed TTTs (James et al., 2020) and underestimating the role of the topography of Madagascar in modelling southern African rainfall (Barimalala et al., 2018). Challenges also arise from the ability of the scientific community to understand and analyse climatic data whether historical or modelled (James et al., 2020).

Conclusion

The climate of southern Africa is highly heterogenous, driven by a range of global to local forcing mechanisms and influenced by the varied latitude, altitude, and distance from the sea. This requires a highly location-specific analysis of any impacts of climate on tourism. Climate change poses a significant threat, more than 1.5 times the global rate of change, with significantly heightened droughts, tropical cyclone risk, and a shift in the biomes which would have an influence on some of the greatest natural attractions in the region. These specific threats of the contemporary and future climates to tourism in southern Africa will be unpacked in Chapter 3. It is important to remember the difference between climate and weather – these climatic patterns determine the range of possible weather conditions in a given place at a particular time in the year. It is ultimately the weather which affects the tourists' experiences, yet the long-term climate influences the tourist seasons, and climate change is increasingly threatening both the climatic suitability of the region and the attractiveness of the destination.

3 Perspectives for approaching climate resource and change in tourism

Introduction

The evolving nature of tourism has interested scholars for almost as long as academic tourism studies have been conducted (see Gilbert, 1939; McMurray, 1930). Change is typical of tourism, and the sector itself represents "a forceful agent of change" (Wall & Mathieson, 2006, p. 6), but it is also a subject of various externally driven changes and impacts at different scales. Currently, one of the key forces of change for the tourism sector and challenge to its future viability is global climate change (Becken et al., 2020; Hall & Higham, 2005; Scott et al., 2019). Climate change, its impacts, and mitigation and/or adaptation needs in tourism represent increasingly important and complicated topics (see Hoogendoorn & Fitchett, 2018; Kaján & Saarinen, 2013; Scott et al., 2012).

According to Scott et al. (2005), research interests focusing on tourism and climate relations started in the 1960s. This so-called formative phase, which lasted until the 1970s, was characterised by interest in general relations between climate and tourism activities (see Paul, 1972). Terminologically, much of the focus was on weather, as climate change was not a topical matter at the time. In practice, however, interlinkages between tourism supply and demand and climate-related elements were realised, but research was not yet able to capture the relationships at a theoretical and/or empirical level. As indicated by Scott et al. (2005), climate issues in tourism were understood to be complicated, but weather and climate data were insufficient for detailed analyses. Furthermore, at the time, most tourism statistics were not available "at appropriate temporal and spatial scales" that would have allowed studies with perspectives on climate-tourism interlinkages (Scott et al., 2012, p. 51).

Internationally, the 1980s was an inactive period in climate- and tourism-related research (Scott et al., 2005). The challenges of inadequate statistical data were still prevalent, and other prolific research topics emerged for both tourism and climate scholars, such as evaluating the socio-economic impacts of rapidly growing tourism (see Franklin & Crang, 2001) and issues with ozone depletion and air pollution, for example (Scott et al., 2012). There

DOI: 10.4324/9781003102618-3

was also limited collaboration between tourism/recreation researchers and weather/climate specialists at the time. Furthermore, this inactive period was probably a result of low awareness concerning anthropogenic climate change and its socio-economic effects on the tourism sector and its future. However, the stagnation period resulted in one of the earliest empirical and highly significant academic studies on tourism and climate change by Wall et al. (1986). They analysed the extended tourism season and the perceptions of visitors to changing weather and climate-related conditions (see also de Freitas, 1990; Smith, 1990). In addition to the changing climate in focus, their study also marks the beginning of scientific adaptation studies in tourism and climate change (Kaján & Saarinen, 2013).

Research interests were re-activated in the 1990s with a focus on the implications of climate change on global travel flows and especially the potential adaptation measures that could be adopted by tourists (see König & Abegg, 1997; Wall & Badke, 1994). According to Scott et al. (2005), this resulting increase in publications by the late 1990s marks the beginning of a maturation stage in tourism and climate change research. A more detailed understanding of supply and demand-side issues in climate change was created, and the sector adaptation actions were actively discussed (see Abegg & Elsasser, 1996; Abegg et al., 1997; Wall, 1998). The maturation stage also resulted in more policy-relevant tourism and climate change research. In this respect, the World Tourism Association's (UNWTO) (2003) Djerba Declaration in Tunis on 'Climate Change and Tourism' initiated a new area of focus within the tourism sector and climate change research. This was followed by the Davos Declaration on 'Tourism and Climate Change' in 2007 (UNWTO, 2007), which called for more efforts on adaptation among tourism businesses and destinations (Scott et al., 2012). Furthermore, it brought forward mitigation issues considering the influence of tourism on climate and the actions that need to be taken to reduce emissions (Kaján & Saarinen, 2013).

The maturation of tourism and climate change research resulted in an increasing academic and policy interest in both adaptation and mitigation issues (Becken, 2005). In the literature, this evolved interest in adaptation, and mitigation has been labelled as a 'house-dividing' process in the research tradition. It led to adaptationist and mitigationist approaches in tourism and climate change studies with partially conflicting interests and goals (Weaver, 2011). In general, the mitigationists are critically concerned about the contribution that the tourism sector is creating for global climate change and how to reduce the emissions based on tourism activities. The adaptationists focus on the mechanisms and perspectives that would help the sector adapt to the impacts of climate change (Hoogendoorn & Fitchett, 2018). If evaluated critically, the mitigationists can be considered to represent a (hard) sustainable development emphasis in tourism-climate change relations, while the adaptationists are working on how to sustain tourism and its future development possibilities in changing climatic conditions. However, this interpretation may over-simplify or misinform us about the adaptationists' missing engagement with sustainable development:

adaptation research is also interested in what ecological, social, and economic elements tourism could sustain and use in a sustainable way in future (see Brown, 2011; Kaján & Saarinen, 2013; Swart & Raes, 2007). Indeed, this kind of division is partly questionable, and some researchers perceived it as just "two sides of the same coin" (Dubois & Ceron, 2006, p. 411). Furthermore, mitigation could be regarded as a tool for a long-term adaptation in which the tourism sector and other key stakeholders proactively adjust their operations within the changing climate policy landscape and emerging emission regulations, for example (see Tervo-Kankare et al., 2018a). However, there are other structural, social, and environmental justice-related issues built into the division of adaption and mitigation needs in tourism, which are highly topical especially in the southern African context.

Global south perspectives

Need for social and environmental justice

Both adaptationist and mitigationist perspectives are actively used in current tourism and climate change research. The former perspective focuses on how the tourism sector, that is, the sector, tourists, and other key stakeholders, aims (or could adapt) to change through transforming their operations or behaviour (Hoogendoorn & Fitchett, 2018a; Kaján & Saarinen, 2013). By contrast, the latter perspective highlights the responsibilities of the sector and the consumers (Saarinen, 2021a) and the need to reduce carbon dioxide emissions and/or enhance carbon sinks by creating and participating in specific compensation programmes in order to reduce the extent of global warming (Scott et al., 2019; Zepper & Beaumont, 2014).

While both approaches are common in tourism and climate change studies, there is a difference in emphasis between the Global North and South. This relates to the specificity of research contexts. In general, high-income countries, regions, or places are considered as the Global North, while the Global South relates to emerging economies or low- and middle-income countries (see Carmody, 2019). This division between the Global South and Global North is not simply a matter of territoriality and bordering in between them, as it also involves complicated issues and differences in power relations, social and environmental justice, inequalities, and unequal division of benefits and costs of global capitalism and tourism (see Jamal & Camargo, 2014; Saarinen & Rogerson, 2021).

Adaptation and mitigation research are both common in the Global North research contexts (Becken & Hay, 2007; Kaján & Saarinen, 2013; Tervo-Kankare et al., 2018a), but mitigation needs, especially, have been highly vocalised in the Global North with respect, for example, to aviation and tourism-related mobilities (see Gössling et al., 2009; Peeters et al., 2019). By contrast, research in the Global South in general, and the southern African scholarship on tourism and climate change, has placed greater emphasis on the adaptationist approach (see Hambira et al., 2013; Hoogendoorn &

Fitchett, 2018a; Saarinen et al., 2020). This difference is grounded in climate justice and environmental and social justice perspectives (see Jourdan & Wertin, 2020). Those are of particular importance for tourism and climate change relations since they are concerned with unequal distribution of possibilities and access for different socio-economic groups and ensuring the equitable distribution of the costs and benefits of (tourism) development. Furthermore, the environmental and social justice perspective is concerned with the mechanisms that empower sustainable development in the Global South and produce injustices in the local global nexus (see Lee & Jamal, 2008; Saarinen, 2021b). In southern Africa, local communities are often dependent on the same natural resources that the tourism sector utilises. This raises critical questions of climate justice, which highlights the actuality that "those least responsible for anthropogenic climate change are often those most affected by its impacts" (Brooks et al., 2020, p. 315).

All this calls for critical analyses on climate change and tourism management and development in the Global South and how that relates to the issues highlighted in the Global North. While climate change mitigation is a global-scale challenge and a highly shared need, an underlining fact is that the continent of Africa, for example, contributes less than 5% of the greenhouse gas emission globally (Khan et al., 2014). Still, the continent, its human population, and the environment carry the burden of the consequences of global climate change impacts that are mainly derived both historically and presently from elsewhere. This unevenness in climate and environmental and social justice between the Global North and South largely explains the lower level of research activity focused on the mitigation questions in southern Africa. Thus, there is a need for research on sustainable adaptation that would seek "to combine the objectives of climate change adaptation and poverty alleviation" (Brown, 2011, p. 23), which makes the research field highly topical in the SDGs policy context.

Furthermore, research in the Global North has a strong focus on the tourism operators and tourists, involving both adaptationist and mitigationist approaches (Scott et al., 2012; Tervo-Kankare et al., 2018a). While tourism business and consumer views are also evident in southern African tourism and climate change research (Dube et al., 2018; Fitchett & Hoogendoorn, 2018; Hambira et al., 2013; Tervo-Kankare et al., 2018b), there has been a notable interest in the community perspectives on climate change impacts on tourism and related livelihoods (Hambira et al., 2021; Hambira & Mbaiwa, 2021). These community aspects and impacts are given much less attention in the Global North (see Kaján & Saarinen, 2013).

Southern African perspectives: towards sustainable tourism-community relations

There is a growing scholarship concerning the effects of climate change on tourism in the southern African region, and countries such as Botswana,

South Africa, Zambia, and Zimbabwe have been increasingly studied in tourism-climate change research (Dube, 2003; Dube & Nhamo, 2019a, 2019b; Fitchett et al., 2016c; Hambira et al., 2021; Moswete & Dube, 2013; Saarinen et al., 2020). In southern Africa, the tourism sector contributes to socio-economic growth and diversification (see Rogerson & Visser, 2020; Saarinen et al., 2020) and serves as a potential instrument for sustainable use of natural and cultural resources (Stone et al., 2020). These aims are highlighted in relation to the development of peripheral areas and rural communities. There are also many different kinds of community-based tourism (CBT) projects and community-based natural resource management (CBNRM) programmes with tourism components (Ngoni & Saarinen, 2021; Roe et al., 2009). These CBT projects and CBNRM programmes emphasise the integral role of local communities in tourism operations and the need to benefit communities in tourism development (Box 3.1).

BOX 3.1 COMMUNITY-BASED TOURISM AND CLIMATE CHANGE IN BOTSWANA

David Lesolle and Naomi Moswete

Tourism plays an important role in rural Botswana as it can provide opportunities for employment, generate income, and activate local businesses (Lenao & Basupi, 2016; Moswete & Thapa, 2018). To develop tourism in rural areas, the Government of Botswana (2007) has initiated CBT. CBT refers to tourism initiatives that are owned by one or more communities or run as joint venture partnerships with the private sector with equitable community participation. CBT is promoted as a means of using the natural (and cultural) resources in a sustainable manner to improve their standard of living in an economically viable way (GoB, 2007). For Botswana, CBT has become a strategy for environmental conservation and community development (Moswete & Thapa, 2018).

Community tourism and climate change

Communities in Botswana are among the most vulnerable to the vagaries of global warming and climate change (Lesolle & Ndzinge, 2018). CBT initiatives are also vulnerable to climate change (Moswete & Thapa, 2018) through a reduction in tourist seasons and limited water availability for the tourists and local people. Reduced water availability due to consumption in the tourism sector can also lead to disputes within host destinations. Thus, extreme weather conditions can negatively affect places that depend heavily on tourism. Erratic rainfall and recurring droughts can impact negatively on the quality of rangelands and existence of wildlife, a necessary precondition for nature-based tourism in Botswana.

Analyses of temperature data demonstrate an increase in the minimum temperatures in Botswana. Even a 1°C increase in mean temperatures can have significant impacts on biodiversity, and thus, CBT operations that are dependent on the attractiveness and stability of environmental conditions if tipping points are exceeded. The geographic spread of malaria mosquitoes has been common in the northern-most regions (e.g. Boteti, Chobe, and Ngamiland), but is now extending into the southern part of Botswana. This is now associated with warming and climate change.

Opportunities for building resilience for CBT in the advent of climate change

While the climate change response agenda can be driven from the top, there is a substantial role that the citizens and communities must take – of their own volition – to make a difference. In this respect, studies have identified the need for climate change awareness for the effective implementation of sustainable community tourism (Saarinen et al., 2020). In most cases, climate change actions to mitigate greenhouse gas emissions and to build climate resilience and climate change adaptation are financeable. CBT is yet to benefit from such opportunities (Mbaiwa & Tshamekang, 2012), and capacity needs to be built around awareness on global warming and climate change.

There is a need to minimise carbon footprint in tourism activities in Botswana across both general tourism and CBT. Tourists would normally have to travel to destinations offering tourism activities. Unfortunately, most travel and transport leads to greenhouse gas emissions. The tourism sector must therefore identify options for minimising the carbon footprint of their activities. The next step would be to offset the carbon emissions associated with community tourism activities, in most cases by taking part in afforestation and reforestation projects. With respect to CBT policy and strategy development, the national strategy that deals with climate change needs to consider both GHG mitigation and climate change adaptation strategies. The strategy must include scheduling CBT activities to avoid extreme weather and climate events and allow for flexible programming of events and by so doing adapting to global warming and climate change.

At the same time, however, the regional tourism sector is noted as highly vulnerable to the estimated impacts of global climate change (Hambira & Mbaiwa, 2021; Hoogendoorn & Fitchett, 2018a). As temperatures in the southern African region are projected to rise faster than the global average in coming decades (James & Washington, 2013; Joshi et al., 2011; Ziervogel et al., 2014), the sector and the tourism-dependent local communities are in a highly vulnerable position. Therefore, an adaptive and climate change resilient tourism sector is needed for sustainable tourism and community relations, which explains the focus on both adaptation and community aspects in tourism and climate change studies in the region. Furthermore, there is a strong development policy emphasis, indicating that the regional tourism sector could and should contribute to the United Nations SDGs (see Rogerson & Baum, 2020), which call for socially informed and responsible approaches in the analysis of the linkages between tourism, climate change, and sustainable development (Scheyvens, 2018). These elements will be discussed later in relation to climate change resilience and policy issues. Next, the importance of climate and weather resources and conditions for tourism are discussed.

Importance of climate (and weather) for tourism

Organising tourist demand and supply in local–global nexus

In the modern history of tourism, suitable climate and weather conditions have played major roles in the evolution of tourism and the success of tourism destination development (Smith, 1998). However, the attractiveness of certain types of climate and weather is historically conditional. This applies to the majority of attraction elements in tourism. For example, the natural environment, wilderness, or wildlife, which are major attractions in the current southern African tourism landscape, received initial positive public interest in the 18th century due to the Romanticism that glorified the past and nature (see Bewell, 2004). Similarly, sun, sea, and sand (3S) tourism that is based on specific warm 'southern' climatic conditions emerged as attractive in the early part of the 19th century based on a "shift in fashion" preferring tanned skin (Scott et al., 2012, p. 54). As a result, the big picture of (international) tourism has been based on the tourist flows from colder or moderate climates with variable and intemperate weather conditions (see Hall & Higham, 2005; Lohmann & Kaim, 1999) towards warmer, drier, and more comfortable climate conditions. However, the specific preferred characteristics of climate and weather are often context-dependent (Jeuring, 2017; Scott et al., 2012). For example, beach tourism benefits from slightly different kinds of climatic conditions than active ecotourism or adventure tourism.

According to Gómez Martín (2005, p. 573), "there is a general tendency among geographers to assume that climate is only important for locating tourism centres when the territorial scale of the phenomenon or the analysis is small (that is, a study of a relatively large area)". This means that climate is seen to define optimal zones for tourism at a global and regional scale, while some other factors are perceived to play a major role in tourism development at a local scale. However, as Gómez Martín (2005) further states, local climatic conditions also influence factors such as the location of resorts, the seasonality, and the spectrum of tourist activities. As a result, many tourist resorts have been able to use local or micro-climatic conditions to their benefit and for product development. Therefore, despite the geographical scale, climate and weather are among the key elements of a destination's geography and how supply and demand are organised in practice (see Becken & Wilson, 2013). At the tourism system level, climate and weather conditions influence tourist perceptions, images, willingness to travel (see Giddy et al., 2017a), and thus, tourist flows and tourism seasons. Obviously, there are many other factors that guide tourism development, but among the environmental elements, climatic issues play a crucial role (Gómez Martín, 2005).

Climatic elements also have major on-site impacts on the local tourism resource base and the kind of products that can be offered and marketed (Scott et al., 2019). Furthermore, they influence visitors' participation in different activities and their overall satisfaction (Becken & Wilson, 2013). Therefore, particular climatic and weather conditions and their changes can support or hinder certain tourism activities (Gómez Martín, 2005) or even lead to trip cancellations in advance (Nilsson & Gössling, 2013). Furthermore, safety issues in tourism are linked to extreme weather events such as storms and heat-waves (Becken & Hay, 2007). Although studies indicate that the tourism sector in southern Africa is aware of the safety issues that weather variability and climate change may create for tourists (see Pandy & Rogerson, 2018; Saarinen et al., 2012; Tervo-Kankare et al., 2018b), there are increasing research needs concerning risks that extreme weather events may create for visitors and infrastructures (see Fitchett et al., 2016a).

Past research has demonstrated the obvious relationship between good (as perceived) climatic conditions and tourists' willingness to travel and their on-site satisfaction. By contrast, poor climate and weather are sources of hesitation in travel decisions, on-site dissatisfaction, and negative experiences (Coghlan & Prideaux, 2009; Gössling et al., 2006; Scott et al., 2012). This link between climatic conditions and overall satisfaction is noted to be potentially more important for destinations that are exposed to highly variable climate and weather (Becken & Wilson, 2013), which highlights the issue of accelerating climate change, increasing intensity, and frequency of extreme weather events and their effects on tourism demand and changing patterns in future supply. Due to their central role, substantial attention has been given to climatic attributes in tourism (Hu & Ritchie, 1993). Based on attributes such as windiness, humidity, temperature, and cloud cover, many attempts have been made to identify the optimal climatic conditions for different tourist activities or tourism destination development in general (see Becker, 1998; Bigano et al., 2006; Lise & Tol, 2002; Hamilton & Lau, 2005; Maddison, 2001). This often has been based on the development of climate tourism/recreation indices (see de Freitas, 1990; Mieczkowski, 1985; Scott et al., 2019) for use by the tourism sector, planners, and policymakers. It is believed that the indices could be useful to tourism planners and the sector for developing and organising future supply and demand in tourism management under changing climatic conditions (see de Freitas et al., 2008). To do so, we need to understand the specific nature of climate as a resource for tourism.

Climate as a resource for tourism

Climate is a resource that the tourism sector uses in its operations. It is often seen as one of the most important resources of a tourist destination (Rutty & Scott, 2010). Climate is a natural resource, but as noted earlier, the meanings and roles of climate and weather are culturally and socially conditional in tourism. The roles of climate involve the so-called push and pull factors in tourism, which form a general framework for tourism mobilities (see Leiper, 1979). This means that it may not be enough that certain sites and regions have pleasant climatic conditions that are, in principle, 'pulling' people to visit those places. In addition, the home environments and climatic conditions of visitors need to differ from those at tourism destinations so that the destination climate becomes a resource. In other words, domestic climate conditions have a capacity to 'push' people to leave home and visit and experience other kinds of climatic conditions than they experience in their everyday lives. In this respect, climate plays a major role in the tourism system.

However, the climate is a complex resource, and it has several specific characteristics in the context of tourism. Gómez Martín (2005) has divided the characteristics of climate as a resource for tourism into six main points:

1   Climate is a free resource. It is a common and abundant resource that needs no mechanism for allocating or sharing it. According to Gómez Martín (2005), no direct conflicts arise from using climate as a resource. However, indirectly, climate and especially its changing characteristics and its local and regional outcomes may create climate-induced conflicts and injustices over the use of water, for example, between different land uses. Golf tourism, for example, is noted to be a relatively water-intensive form of tourism (López-Bonilla et al., 2020; Pandy & Rogerson, 2018) that competes with the resource needs of other local livelihoods.
2   Climate is a resource that cannot be transported or stored. This non-storable characteristic applies to the tourism sector as an experience economy, in general; an unsold hotel room or flight seat cannot

be stored for the next night or flight. Furthermore, in the case of climate, the consumer must travel to a specific place to enjoy the destination and its climatic attributes. Thus, climate-dependent activities are linked to a particular geographical space that has certain atmospheric characteristics.

3 The distribution of the climate as a tourism resource varies in space and time. In practice, this means that there are climatic conditions that limit and/or favour certain tourist activities. Thus, the demand and supply for certain climatic conditions may vary substantially, which also influences the pricing levels (see Scott et al., 2012) between the peak and off-season, for example.

4 As the climate is subject to variation, some related extreme climate and weather events can regionally and locally jeopardise the tourism sector and the existence of its key infrastructure. In southern Africa, these risk events include floods, storms, and heat-waves, for example (Fitchett et al., 2016a, 2016c).

5 Climate is perceived by people in tourism. It impacts tourists as individuals who do not necessarily evaluate the impact identically, but based on a complex set of perceptions that are influenced by individual life courses and social and cultural background. Thus, climate tourism indices need to involve people, that is, consumers, entrepreneurs, and their perceptions and not only climatic attributes and variables.

6 Climate is traditionally seen as a renewable and non-degradable resource. This is based on the simple idea that the climate available in the future is not affected by the amount of it that has been used in the past. However, climate change research demonstrates that contemporary societies, including the tourism sector, contribute increasingly to global climate change and its impacts in various destinations by using the elements and attractiveness of climate resources. This calls for both mitigation and adaptation in tourism development and related studies.

Indeed, the climate is a complex, changing, and fragile resource for tourism. It is often an important part of tourism products, and for Gómez Martín (2005, p. 579), climate is an attraction factor that "acquires greater importance than the other elements" in local and regional tourism products. By contrast to the idea that climate would be a non-degradable resource, however, its future quality, stability, and existence are increasingly threatened by global climate change-triggered by the past and current uses of climate by different economic sectors, including global tourism. For some scholars, this dramatically increased anthropogenic influence on the global climate system has progressed to a level in which humans are regarded as one of the 'great forces of/in nature' (Morton, 2012). In the current environmental literature, this increasing domination by humans is called the Anthropocene (Crutzen, 2002; Latour, 2015), representing a new epoch in which human activity has become the dominant force and influence on climate and the environment in various scales (Rockström & Klum, 2012). As such, it connects global to local: what takes place somewhere in a distant economic production system can have serious impacts on our living environment or preferred tourism destinations (Saarinen, 2019).

Climate change threats to tourism supply, demand, and tourism-dependent communities

> the tourist industry is highly sensitive to weather conditions. It is not known in any detail, however, as to what extent any particular tourist region is affected by the weather in that region or by the weather in the tourists' areas of origin.
>
> (Maunder, 1970, p. 165)

The relationships between climate and tourism are known to be complex. Still, what is increasingly understood is that a changing climate will impact, exacerbate, and create new risks for tourism in the future, especially in the Global South (Hoogendoorn & Fitchett, 2018a; Nyaupane & Chhetri, 2009; Rogerson, 2016). In this respect, global climate change causes impacts and threats to the sector, tourists, tourism-dependent communities, and societies in general. Various impacts and threats are highly interlinked in practice, and they can be direct or indirect by nature. Scott (2019, p. 93) has summarised four pathways through which climate change creates threats for tourism and can alter the prospects for sustainable tourism development: (1) direct changes in climate conditions; (2) indirect climate-induced environmental changes; (3) indirect climate-induced socio-economic changes; and (4) policy responses within tourism and in other sectors. Here, it is important to note that these threats and changes are not shared, that is, experienced, equally in destination regions.

In this respect, local tourism-dependent communities are often the most vulnerable and impacted in the southern African context (Saarinen et al., 2020). This is based on their poor socio-economic conditions and also on their place-based nature: compared to visitors and businesses, communities are not very mobile. Similarly, as climate is a resource that cannot be transported nor relocated, communities are also place-bounded. The visitors are probably the most mobile and flexible stakeholders in tourism; they can constantly change their travel choices and preferences. There have already been indications that tourists react to climate change and its local impacts by changing their travel choices (see Hamilton et al., 2005; Nilsson & Gössling, 2013). Compared to their clients, the tourism operators are less mobile. Still, if environmental (or economic and/or political) conditions become unfavourable for the local and regional tourism sector, the businesses can transfer and relocate their operations into new places, where tourism can be introduced as a prospective tool for local community development.

While there is an irony here, tourism is a labour-intensive sector, and thus, it can be readily introduced to new localities based on its favourable employment effects (Rogerson, 2013; Rogerson & Visser, 2020). This employment effect has been successful in many places. Especially in rural areas and other peripheries, it has transformed many communities to be increasingly dependent on visiting tourists (Saarinen, 2019). In cases where climate change impacts create an unwillingness to visit certain places, adjacent communities are still place-bounded. All they can do is to try to find new alternative livelihoods, but as tourism has often been introduced and represented as the last resort for rural communities, it may be a very challenging task to move on to new socio-economic development avenues. Overall, this limited capacity to adapt makes local communities highly vulnerable to the impacts of climate change in tourism.

In the southern African context, one of the key resources for tourism is water. According to Gössling et al. (2015), total water utilisation in tourism is expected to double over the next 40 years. However, the availability of water is highly dependent on climatic conditions, which are projected to become drier in the coming decades throughout a large part of southern Africa. Intensified droughts have already influenced tourism businesses in the region. One of the most notable recent cases was the Cape Town 'Day Zero' drought spanning 2015–2018. The term 'Day Zero' referred to a date set by the local government, at which point household taps would be switched off, and residents would have to queue at tankers for their daily water allowance (Wolski, 2018). The possibility of water running out completely was a very serious concern and a real possibility, following three anomalously dry rainy seasons, and a drastic reduction in dam levels (Shepherd, 2021). According to Pascale et al. (2020), anthropogenic climate change made this event five to six times more likely relative to the early 20th century.

This drought posed serious threats to the well-developed and mature tourism sector of the Western Cape Province through water restrictions which limited the range of tourist activities and attractions on offer and reduced tourist comfort (Dube et al., 2020; Prinsloo, 2019). Cape Town never reached the 'Day Zero', but as a result of the restrictions, the tourist sector was seriously impacted. For example, there was a clear decrease in bookings, arrivals, tourist spending, and occupancy rates between January 2018 compared to January in 2017 (Cape Town Tourism, 2018; Dube et al., 2020); the occupancy rate in Cape Town hotels, for example, declined by 10%. Longer-term impacts on the impacts of the 'Day Zero' drought for the destination image of Cape Town, particularly for long-haul tourists, remains to be seen (Shepherd, 2021) and has been obscured by COVID-19 lockdown regulations and travel restrictions. These effects of the drought on tourism in Cape Town are of serious concern as the region is projected to experience more frequent droughts of this scale in future decades due to the poleward displacement of the westerlies (IPCC, 2021; Otto et al., 2018; Sousa et al., 2018).

Despite being one single case, the situation in Cape Town with wide media coverage raised awareness of the current and future challenges that climate change can and will create for people and businesses in southern Africa (see Millington & Scheba, 2020). Similar water shortages have been common elsewhere, for example, in the capital cities of Gaborone (Botswana) and Windhoek (Namibia) affecting negatively both residents and visitors. Furthermore, there are many problem regions and places across the region that are very likely to face their own 'Day Zero' in the relatively near future. In general, tourists and their activities require much more water than local residents (Gössling & Peeters, 2015), which may create water use conflicts between the tourism sector and local communities and/or between tourism and other economies and land-use forms (Cole, 2014; Cole & Ferguson, 2015). This will create further challenges for sustainable tourism development in southern Africa in the future.

Conclusions

Climate and weather conditions are some of the most important resources for tourism. Thus, climate change creates challenges for the tourism sector, destination management, and tourism-dependent local communities. The complex relationships between climatic issues and tourism have been known for a long time. In general, the tourism sector is vulnerable to changes in climate and weather. This vulnerability is very evident in the southern African tourism development contexts where regional attraction elements have been largely based on natural landscapes and wildlife, which are increasingly threatened by the impacts of global climate change. As a result, climate change, its impacts and mitigation, and adaptation needs are highly important research topics in southern Africa. Further research is urgently needed as climate change signifies various threats to tourism and its sustainability, which creates social and environmental justice issues and can jeopardise the economic and social viability of the sector. All this can hinder possibilities to use tourism for the achievement of SDGs in the future.

4 Methods for tourism and climate change research in southern Africa

Introduction

A wide range of methodologies have been developed and applied in research on tourism and climate change (Becken, 2013). The selection of methodological approach determines the success in answering the research questions and the stability of the research findings. A strong understanding and description of the chosen methods is fundamental to the replicability of the study and comparability of the results. Both are important components of good science (Broadbent, 2013). Critical engagement with methodologies and innovation towards their improvement and advancement are key components of any discipline. For this reason, methodological reviews have become commonplace in academic journals. Understanding the methodological approaches used within a discipline is an important starting point in considering the outputs of the literature and the grounding they provide for the discipline.

The toolbox of methods for the study of climate change and tourism has, since its inception, drawn from a range of academic fields and has been inherently mixed-method in approach (Figure 4.1). From disciplines of human geography, more qualitative methods for data collection and analysis, including survey questionnaires, interviews, focus groups, and ethnographic studies, have been employed when engaging with governmental stakeholders, tourism operators, and the tourists themselves. From the climate sciences and biometeorology, more quantitative methods have been adopted including trend analysis of meteorological data, climate indices, and SLR projections. The majority of research on tourism and climate change adopts a range of both qualitative and quantitative approaches to determine the threats of climate change to tourism, climatic suitability of the destination, and status of climate resources. This is important in triangulating results and controlling for non-climatic factors.

Southern African research into tourism and climate change has arguably benefited from its relatively late inception, in that much of the foundation of the methodological framework had already been developed. The application of methods directly to a southern African context should, however, be

DOI: 10.4324/9781003102618-4

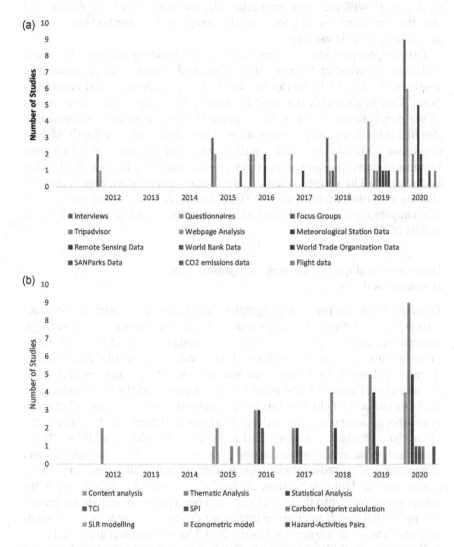

Figure 4.1 Methods of (a) data collection [top panel] and (b) data analysis [bottom panel] in southern African research on tourism and climate change.

exercised with caution due to the unique tourism landscape and climatology of the region. For example, methods developed for a temperate European city may not be appropriate to apply in the context of NBT in the subtropics of southern Africa (Mushawemhuka, 2021; Mushawemhuka et al., 2020). Some methods, although very well suited, may not be possible to apply due to a paucity of data (Fitchett et al., 2016b). For this reason, it is promising that there continues to be considerable development in tourism and climate change methodologies both internationally and within southern Africa, and that there remains a strong focus on the comparison of method outputs and assessment of their validity.

This chapter provides an overview of the methodological approaches and a critical appraisal of their use in tourism and climate change research in southern Africa, highlighting key limitations, challenges, and considerations. This begins with the most frequently used approach – assessments of the perceptions of stakeholders, predominantly operators and tourists, through interviews and questionnaires. This is followed by a South African developed methodology – the use of TripAdvisor reviews as a data source for author-generated content in reviews on, among other factors, the weather and climate. The chapter then explores more quantitative approaches including tourism climate indices (TCIs) and SLR and climate projections. The chapter concludes with analytical approaches used in interpreting the results of these data.

Interviews and questionnaires – engaging with common methods

Drawing from the tourism geographies methodologies which are characterised by more qualitative approaches aimed at exploring the knowledge, perceptions, and perspectives of a sample group (Hambira et al., 2020), interviews and questionnaires have formed the dominant mode of data collection in empirical southern African tourism and climate change research since its inception (Figure 4.1, top panel). A consequence of the interdisciplinary background of the field of study, increasingly studies employing either questionnaires or interviews, or both, as methods to determine the experiences of tourists, tourism operators, and political stakeholders, additionally incorporate more quantitative methods relying on databases compiled from meteorological records, remote sensing imagery, and visitor numbers among others (see e.g. Dube & Nhamo, 2019a, 2020a; Smith & Fitchett, 2020). Investigations solely into the experiences and perceptions of individual groups through either questionnaires or interviews do, however, remain a core component of the literature (see e.g. Giddy, 2019; Hoogendoorn et al., 2021; Pandy & Rogerson, 2019; Saarinen et al., 2020). Key considerations in the use of both interviews and questionnaires include the appropriate sample size and representivity, the mode of delivery, the sampling technique, and the compilation of questions (Table 4.1). Additionally, because human participants are involved, academic institutions may require application to, and clearance granted from, an ethics board including considerations of confidentiality, anonymity, and the recording and storage of data (see Dube et al., 2018).

The sample size in studies using questionnaires is, on the whole, considerably higher than those using interviews (Table 4.1). Southern African studies on tourism and climate change using questionnaires have a mean sample size of 212 respondents, with a range from a minimum of 22 (Giddy

Table 4.1 Sample size, mode of delivery, and sampling technique of interviews and questionnaires exploring tourism and climate change in southern Africa

	Authors	Year	Journal	Sample Size	Mode of Delivery	Sampling Technique
Interviews	Saarinen et al.	2012	Development Southern Africa	7	In-person	Purposive
	*Rogerson & Sims	2012	Urban Forum	10 +?	Not stated	Purposive
	Hambira & Saarinen	2015	Development Southern Africa	9	In-person	Purposive
	*Hambira et al.	2015	Pula: Botswana Journal of African Studies	6	Not stated	Convenience
	*Hoogendoorn et al.	2015	South African Geographical Journal	31	Not stated	Purposive
	Fitchett et al.	2016c	Transactions of the Royal Society of South Africa	75	In-person, door to door	Purposive, convenience
	Fitchett et al.	2016a	South African Journal of Science	53	In-person	Purposive
	Hoogendoorn et al.	2016	Bulletin of Geography	105	In-person	Purposive, convenience
	Mushawemhuka et al.	2018	Bulletin of Geography	21	In-person	Purposive, snowball
	Pandy & Rogerson	2018	Euro Economica	31	Telephonic	Purposive
	Tervo-Kankare et al.	2018	African Geographical Review	9	In-person	Purposive
	*Dube & Nhamo	2019b	Environmental Development	Not specified	Not stated	Purposive
	Mahlangu & Fitchett	2019	Bulletin of Geography	8	Not stated	Purposive, convenience
	Pandy & Rogerson	2019	Urbani Izziv	30	In-person	Purposive
	Dube & Nhamo	2020a	Environmental Development	15	Not stated	Purposive
	Dube & Nhamo	2020b	Journal of Outdoor Recreation and Tourism	50	In-person	Purposive
	*Dube & Nhamo	2020c	The Journal of Transdisciplinary Research in Southern Africa	33	Not stated	Not stated
	Dube et al.	2020	Journal of Outdoor Recreation and Tourism	15	Not stated	Purposive, snowball
	Hambira et al.	2020	African Geographical Review	9	In-person	Purposive
	Pandy & Rogerson	2020	New Directions in South African Tourism Geographies	31	Telephonic	Purposive
	Smith & Fitchett	2020	African Journal of Range and Forage Science	10	Not stated	Purposive, snowball
	Dube & Nhamo	2021	African Geographical Review	88	In-person, door to door	Purposive
	*Hoogendoorn et al.	2021a	African Geographical Review	4	In-person	Purposive

(Continued)

	Authors	Year	Journal	Sample Size	Mode of Delivery	Sampling Technique
Questionnaire	*Rogerson & Sims	2012	Urban Forum	110	Not stated	Not stated
	*Hambira et al.	2015	Pula: Botswana Journal of African Studies	22	Not stated	Convenience
	*Hoogendoorn et al.	2015	South African Geographical Journal	68	Via email	Purposive
	Giddy et al.	2016	Proceedings of the Society of South African Geographers Biennial Conference	57	Via social media	Purposive, snowball
	Giddy et al.	2017a	Tourism Review International	22	Via email	Purposive
	Giddy et al.	2017b	Bulletin of Geography	57	Via social media	Purposive, snowball
	Dube et al.	2018	African Journal of Hospitality, Tourism and Leisure	155	Online and email	Purposive, convenience
	Giddy	2019	International Journal of Event and Festival Management	139	Online and email	Snowball
	Dube & Nhamo	2019a	Environment, Development and Sustainability	369	Online and email	Purposive
	*Dube & Nhamo	2019b	Environmental Development	370	Online and email	Purposive
	*Dube & Nhamo	2020c	The Journal of Transdisciplinary Research in Southern Africa	77	In-person, door to door	Purposive
	Dube & Nhamo	2020d	Bulletin of Geography. Socio-economic Series	370	Online	Purposive
	Friedrich et al.	2020b	Weather, Climate and Society	562	In-person on-site	Purposive, convenience
	Friedrich et al.	2020a	Transactions of the Royal Society of South Africa	562	In-person on-site	Purposive, convenience
	Saarinen et al.	2020	Development Southern Africa	289	In-person, door to door	Purposive
	*Hoogendoorn et al.	2021a	African Geographical Review	170	On-site, in-person	Convenience

et al., 2017a; Hambira et al., 2015) to a maximum of 562 (Friedrich et al., 2020a, 2020b). For interviews, the mean sample size is 30 ranging from 4 (Hoogendoorn et al., 2021) to 105 (Hoogendoorn et al., 2016). In some instances (Dube & Nhamo, 2019b; Rogerson & Sims, 2012), the total number of persons interviewed is not reported. It is clear from these statistics that there is no agreed minimum threshold for the sample size; rather, the way in which the data and any assumptions regarding representivity are handled. In this sense, the discipline is not reliant on quota sampling (Trochim, 2008). Indeed, many of the papers on tourism and climate change for southern Africa explicitly state that the results presented cannot be deemed representative of any broader community, and that further quantitative research would be required to achieve this (see e.g. Saarinen et al., 2020). In some settings, a response rate can be determined from a total number of potential participants. This is often the case when interviewing or administering questionnaires to tourism operators, where a full listing can be obtained from tourism boards or counted manually (Dube & Nhamo, 2020c; Fitchett et al., 2016a; Giddy et al., 2017a; Mahlangu & Fitchett, 2019). In these instances, researchers can approach the full identified population and report a response rate from those who agree to be interviewed or complete the questionnaire. This meets the criteria of probabilistic sampling (Trochim, 2008).

When engaging with tourists, residents of an area, or working in large metropolitans, it is seldom possible to determine what the full sample size would be (Saarinen et al., 2020). For example, when interviewing tourists on a beach with a large catchment area on a particular day or week, it is not possible to determine the proportion of a total population of beach-going tourists that they represent. In these instances, nonprobability sampling is employed, and a randomly selected population forming a large sample size is preferable (Trochim, 2008) to ensure a degree of redundancy in the patterns emerging from the data (Friedrich et al., 2020a, 2020b). On the opposite side of the spectrum, when engaging with expert sources or key informants, a very small sample size is appropriate as any larger group would involve colleagues or subordinates who would likely provide very similar responses, and thus redundancy is quickly reached (Dube et al., 2021a; Hambira, 2018; Smith & Fitchett, 2020).

BOX 4.1 QUESTIONNAIRES IN SOUTHERN AFRICAN TOURISM AND CLIMATE CHANGE RESEARCH

Jonathan Friedrich

Questionnaires allow for collecting and analysing standardised quantitative data on a specific topic. This box reflects on some issues and limitations associated with questionnaire-based data collection and analysis in southern African climate change and tourism research, grounded in experiences in the fieldwork for recently published studies which explored climate perceptions of beach tourists in South Africa (see Friedrich et al., 2020a, 2020b, 2021). These thoughts can be taken into account when conducting quantitative empirical research on climate change and tourism in South Africa.

Questionnaire-based data collection is often based on secondary data for *a priori* sample definition or sample stratification. Statistics South Africa offers data on tourist travel, in terms of inbound and domestic travel, and expenditure on the national level. The data quality provided by official sources relating to the frequenting of touristic attractions such as beaches and touristic destinations on a regional or local level is rather narrow, which does not allow the delimitation of *a priori* sample size for, for example, sample stratification in relation to previously researched destinations and their visitors. This data availability must be taken in account when planning and conducting empirical research.

When exploring climate perceptions of tourists, a key issue exists in sampling. Tourists whose climatic preferences are not met during the period of field work are unlikely to be outdoors at attractions such as beaches and will therefore be under-represented in the sample. This was experienced during fieldwork in Gqeberha, where the weather was very windy at the time. This meant that less people were visiting the beach, and only a few agreed to participate in the survey due to the adverse weather. Financial and time restrictions when conducting fieldwork can also result in smaller sample sizes anticipated. Conducting research with potential visitors before they are on holiday and while they are still at home can complement the findings of our studies and overcome the described bias, but would not provide an experience-based account.

Tourist participation in questionnaires is another key determinant of the success of the study. Tourists, and especially beach tourists, are on holiday and hoping to relax. As a result, some tourists are reluctant to spend their vacation time filling out questionnaires. This can be compounded by the length of the questionnaire. A short questionnaire that requires to 5–10 minutes to complete is likely to attract a larger sample size. However, the brevity of the questionnaire leads to difficulties in terms of collecting in-depth data which is necessary to get a comprehensive account of tourists' perceptions, as this would require more questions. In addition, beach tourists were defined as people either sitting, walking, or lying at the beach. This means that potential participants who were swimming in the sea (or partaking in other activities) at the time of our fieldwork may not be represented either and could have a different profile in terms of climate sensitivity. The South African summer holiday season attracts many local and international beach tourists. This forms a prime period for successful data collection. However, if the researcher only conducts field work in the peak season, tourists visiting the beaches in the off-season are not represented.

Research on perceptions of weather and climate often includes questions that deal with hypothetical scenarios. For example, beach tourists may not have necessarily experienced flooding or have a first-hand understanding of the impact of a tropical cyclone on their travel plans. This particularly relates to extreme weather events that are projected to occur more frequently under climate change. Thus, questionnaire-based data can fall short in this respect and could be complemented with qualitative interviews that deeply interrogate the individual perceptions and phenomena such as risk perception paradox (see e.g. Wachinger et al. (2013) for more information).

Future studies may take into account the limitations experienced in questionnaire-based research in southern Africa including the studies described here. It is important that findings are triangulated using different methodological approaches such as qualitative interviews, data collection in the off-season, or data collection from tourists when they are not on holiday. This will ideally help to strengthen the knowledge on beach tourism and climate change in southern Africa and support both the government and tourism sector with the knowledge to define and apply sufficient adaption strategies and policies to cope with on-going climate change.

Purposive sampling is the most commonly employed approach used in studies on tourism and climate change in southern Africa (Figure 4.1), albeit this is often inferred from the methodological description rather than explicitly stated. In some instances, the study location such as a particular beach (Friedrich et al., 2020b), pre-selected ward (Saarinen et al., 2012; 2020), or a particular Facebook group (Dube & Nhamo, 2019a) is purposively selected, but the pool of respondents is determined via convenience sampling. Similarly, when interviewing stakeholders, an initial cohort may be purposively sampled, and from those respondents, a process of snowball sampling is initiated (Smith & Fitchett, 2020). Arguably, in these instances, where the identified group to be interviewed is determined by their expertise, position, or managerial role (see Dube et al., 2021a; Hoogendoorn et al., 2021; Smith & Fitchett, 2020), the term purposive sampling may be more accurately replaced by expert sampling (Trochim, 2008).

Interviews remain most commonly conducted in-person, although in almost half of the papers using interviews, the mode of engagement is not stated (Table 4.1). Only two of the papers indicate that they conducted interviews telephonically. More recent and forthcoming research which involved data collection during 2020 and future years will likely demonstrate a shift to remote modes of communication due to COVID-19 lockdowns both in southern Africa and globally (Scerri et al., 2020). For questionnaires, remote methods were already commonplace: 9 of the 16 papers on tourism and climate change in southern Africa that used questionnaires administered these online either through email or social media or both. Only five papers disclosed that the questionnaires were administered to people in-person. Social media is increasingly allowing for random sampling and often allows distribution of questionnaires to a much larger potential audience than in-person or email approaches would facilitate (Dube & Nhamo, 2019a). Emails do, however, present two advantages. First, they allow for a known sample number of potential respondents to be approached, each of whom meet some predefined criteria (Giddy et al., 2017a). Second, follow-up requests for responses and to engage further with respondents are more easily facilitated (Hoogendoorn et al., 2015).

BOX 4.2 INTERVIEWS IN SOUTHERN AFRICAN TOURISM AND CLIMATE CHANGE RESEARCH

William Mushawemhuka

Interviews are an important qualitative data collection method frequently used in social science research. Tourism research often makes use of interviews, particularly semi-structured in-depth interviews. In southern Africa, there has been a steady increase in tourism and climate change research, much of which uses in-depth semi-structured interviews (Hoogendoorn & Fitchett, 2018a). These interviews are powerful as a stand-alone research method, and they can also be used to complement other research methods in a mixed-method approach (Dube & Nhamo, 2020b). Within tourism and climate change research, interviews have been used to address research biases and gaps that normally arise from quantitative research. These include aspects such as human perceptions and attitudes towards weather-related changes and extreme weather events as well as the impacts of these changes on communities (Smith & Fitchett, 2020).

As tourism and climate change research evolves, it is proving to be more complex, given the new scientific evidence being presented by contemporary technological advancements. This now warrants mixed-methods approaches which accommodate qualitative data and the numerous contemporary quantitative data collection methods

(Smith & Fitchett, 2020). In contemporary tourism research, interviews are proving to be a useful tool for providing the in-depth com-
prehension of an area or phenomena, where differences in practice,
behaviours, impacts, attitudes, and perceptions may exist. The wide
use of semi-structured interviews in southern Africa has facilitated an
increase in tourism and climate change research in the region over the
past three decades (Rogerson & Visser, 2007). These have had a distinct advantage in the region, as they aid in explaining context-specific
phenomena from under-researched areas such as Zimbabwe, Lesotho,
Namibia, and Botswana (Hambira et al., 2020; Hoogendoorn et al.,
2021; Mushawemhuka, 2021). Despite these aforementioned advantages, the use of interviews as a data collection method can also be
fraught with challenges. These challenges can arise from a lack of access to participants and resources or from the research context itself.

Even though there is a considerable amount of tourism research that
has been carried out in Zimbabwe, there are a number of challenges
that have been faced by researchers as a result of the socio-economic
and political context of the country (Dube & Nhamo, 2019a, 2020b;
Mushawemhuka, 2021; Mushawemhuka et al., 2018). A good starting
point for tourism research would be getting a database of the tourism
stakeholders from local authorities and attaining an insight into the
government and local authorities' strategies towards climate change.
Unfortunately, this information is not readily available, and most government and local authorities are reluctant to be interviewed. Trying
to access information via the telephone or emails from the same of-
fices is equally problematic. In most cases, there is no response at all
(Mushawemhuka, 2021; Mushawemhuka et al., 2018).

An alternative to this challenge would be to obtain information
from private tourism stakeholders. However, when approached, they
immediately ask if you are politically affiliated, with most expressing
that their businesses are not affiliated to any political party, highlighting a sense of fear. In most cases, they too also rarely respond to emails
or consider telephonic interviews, as they perceive the interviews to
be of no financial benefit to them. Despite this, private stakeholders
within the tourism sector are generally more receptive to academic
research, especially if the researcher is physically present, and carrying out interviews in the premises of private stakeholders which has
proved to be successful in most areas around Zimbabwe. Physically
visiting establishments and setting up in-person meetings has yielded
more positive results in Zimbabwe. However, another challenge in
this regard is the dilapidated road infrastructure in Zimbabwe, which
makes access to most tourism destinations around the country very
difficult. This is worsened in the rainy season during which the road
infrastructure is in its worst condition. In addition, travelling around
the country is further inhibited due to fuel shortages that affect the

country, making mobility very difficult. In both 2016 and 2019, I faced this situation while doing fieldwork in Zimbabwe and had to wait in fuel queues for hours on end, which delayed the data collection process of the research.

Despite these numerous challenges faced when carrying out interviews in Zimbabwe, data collected from interviews is very valuable to qualitative research outputs, particularly in under-researched areas such as Zimbabwe. When carrying out tourism and climate change research, interviews present numerous benefits and are more beneficial when merged with various quantitative models, which then presents a well-rounded comprehensive study (Mushawemhuka, 2021). These challenges that are faced in Zimbabwe are most likely to be experienced in other parts of southern Africa since countries in this region share relatively similar socio-economic, political, and geographic settings. COVID-19 has crippled most travel activities globally, affecting both the number of tourist operations and tourists to be interviewed and restricting access to destinations to conduct in-person interviews.

Once data have been acquired via questionnaires or interviews, they are transcribed before analysis. Where online survey platforms such as SurveyPlanet, SurveyMonkey, or GoogleForms are used, the transcription occurs automatically. The majority of studies employ a combination of thematic analysis, content analysis, and basic statistical analysis (Figure 4.1). The distinction between content and thematic analysis is seldom clearly articulated in studies of tourism and climate change nor indeed more broadly (Vaismoradi et al., 2013). Nonetheless, both pre-determined or deductive and emergent or inductive themes from interview and questionnaire data provide valuable data (Walters, 2016), particularly in exploring tourists' and tourism operators' perceptions (Hoogendoorn et al., 2016) and adaptation strategies (Hoogendoorn et al., 2021; Pandy & Rogerson, 2019) relating to climate change. Statistical analyses include very basic counts, frequencies, and proportions (see Giddy, 2019; Giddy et al., 2017b) and more complex approaches such as correlation and cross-tabulation (Friedrich et al., 2020a) and factor analysis (Friedrich et al., 2021). To couple qualitative and quantitative data, studies include both ad-hoc mixed-methods which present each result in sequence (see Fitchett et al., 2016a), and more recently, emerging integrative approaches tailored to specific questions such as the Hazards-Activities Pairs (Friedrich et al., 2020b).

Alternatives to questionnaires – TripAdvisor as a data source

In completing questionnaires, and to a lesser extent interviews, respondents are guided by wording and structure of the questions asked, which can prompt them to consider a factor in greater detail than they would ordinarily constrain their responses to pre-determined classes or lead a particular answer (Driscoll, 2011; Fitchett et al., 2020). The positionality of the researcher may also influence the selection of willing participants and may influence the responses obtained (Hoogendoorn & Visser, 2012; Prinsloo, 2019). Denstadli et al. (2011) emphasise the importance of self-reported satisfaction in assessing climatic suitability of a region for tourism and in understanding tourists' short-term sensations of and adaptation to weather. TripAdvisor reviews provide a platform for this to be achieved (Fitchett et al., 2020), and the analysis thereof is an application of the emerging field of netnographic research in science (Amaral et al., 2014; Jeacle & Carter, 2011).

Reviews on TripAdvisor are user-driven, opt-in, and free-form accounts of tourists' experiences of their vacation, accommodation, and activities (Fitchett & Hoogendoorn, 2019; Fitchett et al., 2020). Consequently, only where weather or climatic factors form a memorable or significant aspect of a vacation or experience would a tourist choose to write a review that includes these factors (Fitchett et al., 2020). The reviews can therefore be seen as a more objective measure of the overall sensitivity of tourists to climate at a particular destination, and of the climatic factors which are most important to the tourists (Stockigt et al., 2018), against which more quantitative measures of the climate itself can be tested (Fitchett & Hoogendoorn, 2019). Perhaps, most significant is that the analysis of TripAdvisor reviews represents the first methodological development in tourism and climate change research which has emerged from the southern African region (Fitchett & Hoogendoorn, 2018, 2019).

The ability to achieve pure objectivity has been challenged in variety of research fields largely due to the reflexiveness of the research process that straddles both subjectivity and objectivity (Poon & Cheon, 2009). Broadbent's (2013) notion of stability should also be considered – as online platforms rapidly change in their structure and as tourist visitations fluctuate for a range of reasons outside of climate, the reproduction of the study with a newer set of TripAdvisor reviews may well produce quite different results.

The methodology for the assessment of climatic sensitivity from TripAdvisor reviews (Fitchett & Hoogendoorn, 2018, 2019; Fitchett et al., 2020) involves the manual reading and coding of a significant sample of TripAdvisor reviews for a particular destination or set of destinations. This ranges from the total of 259 reviews that had been written for Afriski (Stockigt et al., 2018) to over 5,000 reviews for South Africa (Fitchett & Hoogendoorn, 2018, 2019) and over 13,000 reviews for the Indian Ocean Islands (Fitchett et al., 2020). The coding includes whether climatic variables have been mentioned in the review and the specific climatic variables mentioned. Additional information from the TripAdvisor platform including the date of the review and any comments on date of travel and the country of residence of the reviewer are also captured. It is important to collect data spanning a full calendar year, as there are considerable inter-annual variations in the number of TripAdvisor reviews, the proportions of reviews mentioning climate, and the climatic factors cited (Fitchett & Hoogendoorn, 2018; Stockigt et al., 2018). Analysis of the coded database then includes the proportion of climatic mentions, frequency distributions of the climatic factors, seasonal distributions of reviews and climatic factors, classification of reviews by country of origin, location visited, and quality of accommodation establishment, sentiment analysis of the commentary on climatic factors, coupled with quotes from the reviews to illustrate the findings (Fitchett et al., 2020). Principal Component Analysis has additionally been used to explore the climatic factors most salient per destination (Fitchett & Hoogendoorn, 2019). This has set up a framework to not only explore the climatic sensitivity of tourists and their self-reported experience of the weather but also confirm the outputs of quantitative assessments of meteorological data.

TripAdvisor reviews have also been used in a more superficial manner to obtain a broad understanding of tourists concerns regarding climate change at Victoria Falls (Dube & Nhamo, 2020d). While 5,095 of the over 60,000 reviews on Victoria Falls are considered, quantification relating only to the user review ratings of the attraction is presented. While comment is made that tourists are concerned about the waterfall drying up (Dube & Nhamo, 2020d), the proportion of reviews that this represents is not reported. The key finding of this particular analysis of TripAdvisor reviews – that the waterfall is drying up due to a drought, and that this represents an instance of last chance tourism, have both been contested in more recent papers (Hoogendoorn, 2021; Mushawemhuka et al., 2021). This does not, however, discredit TripAdvisor as a source of data, but rather raises caution regarding the methodologies adopted in exploring TripAdvisor reviews, and the need for academic rigour (Landers et al., 2016).

There is considerable scope for further development of TripAdvisor as a data source and of the methodologies for analysis with the continued development of machine learning and artificial intelligence (Karthikeyan et al., 2019). Most important would be the reduction of the time commitment and margin for human error in manual reading and coding of the often >5,000 reviews (Landers et al., 2016). Webscraping tools are becoming increasingly prevalent in biometeorological and broader climate change research, most often applied to Twitter (see Al-Saqaf & Berglez, 2019; Cecinati et al., 2019). Exciting developments in this space have included the development of code to classify emotions and sentiment (Sailunaz & Alhajj, 2019) including webscraping on the TripAdvisor platform (Agüero-Torales et al., 2019; Valdivia et al., 2017). Most importantly, webscraping allows for the number of reviews analysed to increase significantly, improving the degree of representivity of the results (Banerjee & Chua, 2016). Preliminary feasibility assessments in the tourism and climate change domain have, however, revealed the need for deliberate research efforts into effective coding, validated through cross-comparison to manually analysed datasets, particularly to accurately handle terms with double meanings (Pillay, 2019). These include, for example, a cold day versus a cold stare or a warm breeze versus a warm welcome (Pillay, 2019).

Tourism climate indices

Tourism climate indices (TCIs) refer broadly to a growing collection of indices which quantify the climatic suitability of a destination on the basis of meteorological data. The original TCI was developed by Mieczkowski (1985) to assess the suitability of climate for world tourism on the basis of expert opinions. Perch-Nielsen et al. (2010) refined this index to use daily rather than monthly resolution meteorological data and to reduce the number of variables to those which are commonly recorded. Following this, a proliferation of indices have emerged, many of which are tailored to specific tourism attractions including beach tourism (Rutty et al., 2020), urban tourism (Scott et al., 2016c), and camping (Ma et al., 2020). Increasingly, these are developed on the basis of tourists' climatic preferences as evidenced from questionnaire responses rather than expert opinion in the selection and weighting of variables (Scott et al., 2016c). While this does refine the index, it becomes inherently specific to the region in which tourists were surveyed.

Each of these indices produces a score out of a maximum of 100, with a rating scale to classify the suitability of a particular destination for tourism. These classifications usually range from unsuitable on the lower end through to excellent or ideal on the high end. Within these categories, more subtle differences between destinations can be discerned from the index score itself. The scores are calculated to a monthly resolution, thus allowing for the peak seasonal climatic suitability to be assessed according to Scott and McBoyle' (2001) six seasonal distribution classes. Change in mean annual, seasonal, or monthly TCI scores can also be assessed through time (Fitchett et al. 2017; Mushawemhuka et al., 2020; Perch-Nielsen et al., 2020).

BOX 4.3 TOURISM CLIMATE INDICES

Kirsten Noome

Climate is a key component in the process of selecting the destination, timing, and duration of the trip (Gössling & Hall, 2006). Many methods have been proposed to prioritise climate information for more advanced decision-making to maximise tourism opportunities. Climate indices are multi-faceted mathematical models combining tourism and climate data to provide quantitative assessments of tourism resources affecting destination suitability (Matthews et al., 2021). There has been a proliferation of indices over the last two decades, all derived in some form from the original TCI.

TCIs have been widely applied at different levels and in various geographical settings to assess the potential impacts of changing climatic resources. Mieczkowski (1985) first identified the need for an index that evaluates the climatic conditions of destinations for tourists, developing the TCI. This index combined and scored seven climatic components that were determined through expert opinion according to their suitability for thermal comfort through biomedical research (Mieczkowski, 1985). The TCI was devised to be accessible and widely applicable, using commonly available data and allowing the comparison of scores between regions (Fang & Yin, 2015).

There are inherent limitations of the TCI. The index does not take into account the overriding effects of rain and wind; mean monthly meteorological data are used, dampening the effects of daily weather conditions; the index has arguably not been empirically tested globally; and the index is not specific to individual types of tourism (Scott et al., 2016c). Perch-Nielsen et al. (2010) addressed many of these limitations through three adjustments to the original TCI equation and input variables. These include the use of daily meteorological records, adjusting the thermal rating to use effective rather than apparent temperature and replacing the wind-chill index with wind-chill equivalent temperature. More recent efforts have been made to validate the TCI against empirical results, comparing the sensitivity to that self-reported by tourists in TripAdvisor reviews (Fitchett & Hoogendoorn, 2019).

Concurrent to this have been more extensive overhauls to the index. The Climate Index for Tourism (CIT) was developed to modify the TCI for sun, sea, and sand tourism (de Freitas et al., 2008). Rather than build on the TCI, this was a new index, developed through examining tourist climate preferences and key thresholds of daily temperature, sunshine, wind speed, and the absence of rainfall (de Freitas et al., 2008). The CIT recognises the overriding effects of physical weather. For example, regardless of the temperature, high wind speeds and rainfall will cause beach tourists to leave the beach due to the unpleasant weather. The Holiday Climate Index (HCI)$_{Urban}$, developed by Scott et al. (2016c), claims to be the most robust representation for urban tourism. The HCI$_{Urban}$ was designed for leisure tourism and specifically to address the major criticisms of the TCI. The main advancement, according to the developers of the index, is that it is empirically validated in the tourist marketplace (Scott et al., 2016c). However, while HCI$_{Urban}$ and HCI$_{Beach}$ have been developed and validated against European urban and beach tourist preferences, the TCI has been used and tested in a more diverse range of destinations and attractions. While generic indices, such as the TCI, allow for global comparison of climatic suitability, across a range of tourist attractions, a specialised index such as the HCI$_{Urban}$ or Camping Climate Index (CCI) would arguably facilitate more sensitive comparison within a group of attractions.

The TCI has been the most extensively used tourism climate index; yet, in 2018, the number of papers using the HCI in a given year surpassed that of the TCI. It is notable that a significant proportion of those papers using the HCI were authored by one or more of the researchers who developed this index. It remains to be seen whether this index will be adopted more widely, and the reduction in use in 2019 and 2020 is notable in this regard. The TCI was the first tourism and climate-based index to be applied in southern Africa (Fitchett et al., 2017). It has since been applied to three neighbouring southern African countries, namely Lesotho, Namibia, and Zimbabwe (Mushawemhuka et al., 2020; Noome & Fitchett, 2019, 2021). In comparing the outputs of the TCI and HCI for Zimbabwe, Mushawemhuka et al. (2020) report an inflation in HCI scores due to the negating of night-time thermal comfort.

Depending on the purpose of the study and the region of interest, each of these indices has the potential to be the most appropriate. For example, when comparing exclusively camping destinations, the CCI would be more appropriate than the more generic TCI, and while some camping destinations are in beach settings, the BCI or HCI$_{Beach}$ would be too specific for comparison. Where no specific index has been developed, such as the case for adventure tourism, a more widely applicable index would be more appropriate. Rather than advocating for a single index to be used universally, the existing suite of indices each has value when used appropriately for the context (Figure 4.3.1).

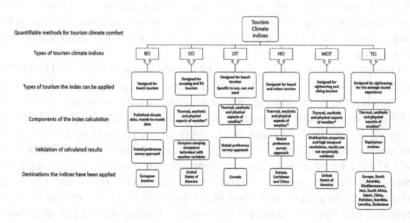

Figure 4.3.1 Flow chart detailing the different tourism–climate indices, a matrix to determine which index to use depending on environment, climatic variables, and country of interest.

Mieczkowski (1985) included Africa in the global output of the TCI at coarse resolution. In southern Africa, Perch-Nielsen et al.'s (2010) adaptation of Mieczkowski's (1985) TCI has been applied to 19 destinations in South Africa (Fitchett et al., 2016a, 2017), Afriski in Lesotho (Noome & Fitchett, 2019), and seven locations in Namibia (Noome & Fitchett, 2021). The weightings of climate factors and rankings of stations from this version of the TCI have been validated against TripAdvisor reviews for South Africa, with suggestions for minor adjustment to the weighting of wind (Fitchett & Hoogendoorn, 2019) and aligns for Lesotho (Noome & Fitchett, 2019; Stockigt et al., 2018). The TCI has been further adapted for southern Africa, where sunshine data are seldom available, through a cross-comparison of the full and adapted index in South Africa (Fitchett et al., 2016b), and applied to Zimbabwe (Mushawemhuka et al., 2020). While inter-model comparisons have demonstrated that the Holiday Climate Index (HCI) is arguably more suitable than the TCI in Europe (Scott et al., 2016d) and the Caribbean (Rutty et al., 2020), the primary adaptation involves the removal of night-time thermal comfort with the argument that in the 30 years since the development of the TCI most accommodation establishments have had air-conditioning installed. This is, of course, not the case for much of southern Africa (Mushawemhuka et al., 2020), and indeed, heat is frequently cited in TripAdvisor reviews of South Africa (Fitchett & Hoogendoorn, 2018) and the Indian Ocean Islands (Fitchett et al., 2020), particularly in the lower-rated accommodation establishments and regions which experience frequent power outages. A similar finding has been reported in an inter-comparison of the two indices for Iran, where the TCI was reported to be more representative of the attractiveness of the region for tourism (Hejazizadeh et al., 2019). Moreover, the HCI was developed in the context of urban tourism in Europe and later refined for beach tourism; the TCI remains the only index to date which was developed with explicit consideration of the African climate and of 'game viewing in African national parks' (Mieczkowski, 1985, p. 225), and which has been verified for this region.

Going forward, it will be of value to test the broader range of TCIs to the locations in southern Africa for which their *a priori* assumptions hold – that is, the HCI for urban settings where air-conditioning is more common, the CCI (Ma et al., 2020) for destinations which offer camping, and the Beach Climate Index (BCI; Morgan et al., 2000) or HCI_{Beach} (Rutty et al., 2020) for coastal destinations. We contend that each of these indices offers value and should be applied to the environment for which it is developed and encourage further index development for southern Africa, and the range of common attractions including beach tourism, NBT, adventure tourism, business tourism, and cultural tourism. Following the lead of the World Meteorological Organization's Expert Team on Climate Change Detection and Indices, an objective assessment of all indices (Donat et al., 2013), and a formal listing of the numerous applicable indices to sectors and climatic conditions would be of value towards progress in tourism and climate research.

Sea level rise and climate projections

While not methodologies specific to, or developed for, tourism and climate change research, SLR modelling and climate projections form important methodologies often used within the subdiscipline. In many cases, researchers on tourism and climate change rely on SLR or climate projection products that are in the public domain either in their own analysis or in the discussion rather than performing the modelling themselves (see e.g. Dube & Nhamo, 2019a; Friedrich et al., 2020b). Here, the quality of the results is only as strong as the accuracy of the projections used, and the extent to which these are effectively downscaled to the southern African atmospheric circulation, topography, and land cover (James et al., 2018). Many papers presenting quantitative analyses of climate data over a period of three or more decades can present tentative projections of future conditions, assuming the trends over those decades persist, and under the condition that they are statistically significant and appropriately calculated (see e.g. Dube & Nhamo, 2019a; Fitchett et al., 2017). This assumes linearity in the time trends, which broadly holds for short time periods, but will become less accurate further into the future. The only study, which has included independent sea level rise projections (Fitchett et al., 2016a), uses a very basic bath-tub approach, with a relatively low spatial resolution. Recent developments in sea level rise modelling (see Watson, 2017) and continual improvements in climate modelling as captured by the IPCC (Flato et al., 2013; IPCC, 2021) have the potential to contribute to more specific and accurate projections of future tourism climates. These would benefit from greater collaboration between climate modellers and those working in tourism and climate change in southern Africa. Using climate model outputs for the variables required to run the TCI has facilitated valuable directly calculated future projections of climate suitability for tourism in Iran rather than inferences from time series (Roshan et al., 2016). These specifically tailored climate change products, and the interpretations of researchers working with both qualitative and quantitative data, may be of particular value to tourists, tourist accommodation and attraction operators, and insurance companies as they each plan for the future (Fitchett et al., 2016c; Olya et al., 2019; Thistlethwaite et al., 2018).

Conclusion

Research into tourism and climate change in southern Africa has benefited from a toolbox of methodologies both from the pre-existing established regional disciplines of tourism geographies and climate sciences and from the international literature on tourism and climate change. This has allowed for the results of studies in southern Africa to contribute effectively to the international discourse. It is also significant that within its infancy, southern Africa has contributed to methodological developments within tourism and climate research through the framework for the analysis of TripAdvisor reviews. As the discipline is growing more rapidly, and the authorship is diversifying, further innovations from the region are likely to be forthcoming.

As an inherently interdisciplinary field, mixed-method approaches and the triangulation of results are critical to robust research. The common integration of qualitative and quantitative databases is thus very appropriate in this setting, as are approaches such as the Hazards-Activities Pairs analytical framework in analysing the results. The specificity of climate threats to tourism by location, tourist attraction, quality of accommodation, and the country of origin of tourists also calls for a suite of methods that can be tailored to the particular setting and research question. For this reason, we encourage a larger rather than more restrictive suite of methods, each of which is tested against tourist experience data. That said, where there are appropriate methods endorsed by, for example, the World Meteorological Organization, these should be used as at least part of the methodological approach, particularly in place of more superficial ad-hoc approaches.

Perhaps, most important is the explicit statement of study methods to enable the reproducibility of the research. The critical review of the existing literature on tourism and climate change highlights the degree to which methods regarding sampling approach, sample size, data checking and homogenisation, and approaches to the analysis of data are seldom stated. Improving on methodological reporting, and the selection of methods to enable triangulation of results, should be key aims of future tourism and climate change research.

5 Climate change risks to southern African tourism

Introduction

Southern Africa has considerable climatic heterogeneity, spanning tropical conditions in Mozambique to arid conditions in Namibia, and frequent snowfall in Lesotho (Engelbrecht & Engelbrecht, 2016). Tourists' preferences are likewise highly varied, depending on the primary attraction they are visiting, their origin, and their expectation of the weather during their vacation, which determines the clothing they will pack (Fitchett & Hoogendoorn, 2019). Tourist preferences also vary from destination to destination across the globe (Scott et al., 2016c). This means that a delicate balance exists between climate resources and the tourism attractions that have developed strategically to maximise the benefits of the weather and climate (Scott et al., 2012). Therefore, any change in climate is of concern, with the potential to threaten this balance, and the impacts of climate change and risks to tourism will be specific to the location and type of tourist attraction (Gómez Martín, 2005). Different types of tourism – nature-based, coastal, cultural, heritage or arid tourism – will be impacted by different climate threats and with distinct levels of intensity and severity depending on the weather and climatic conditions of the location and the climatic requirements and sensitivity of the attraction (Fitchett & Hoogendoorn, 2018). These risks include, but are not limited to, widespread drought, heat, changes in the intensity of rain, flooding, and SRL, which, in turn, heightens the difficulty with which the tourism sector can provide its key services, products, and experiences to tourists (Becken & Hay, 2007).

The way in which these climate change impacts will play out in the southern African tourism landscape is, to a large extent, unclear to tourist operators (Dube et al., 2021a; Gössling & Hall, 2006; Hoogendoorn et al., 2016). A growing body of scientific literature is, however, beginning to document the effects of contemporary climate change-induced extreme events for particular regions or tourism attractions (see e.g. Dube et al., 2020; Fitchett et al., 2016c; Giddy et al., 2017a; Smith & Fitchett, 2020) and to project future impacts for the tourism sector (see Dube & Nhamo, 2019b, 2020a; Fitchett et al., 2016a). A larger body of literature presents the climatological and

DOI: 10.4324/9781003102618-5

ecological evidence for contemporary and future climate change for southern Africa, while the well-established tourism discourse outlines the distribution of tourism resources across the region. This chapter brings together these three key sources to outline the current challenges and likely future risks of climate change to the southern Africa tourism sector.

Threats of gradual climatic change

The southern African region has been warming at 1.5 times the global average, with a projected increase in mean temperatures for the region of 4°C–6°C by 2100 depending on the level of reduction in carbon emissions in the near future (Archer et al., 2018; IPCC, 2021; Ziervogel et. al, 2014). This change in mean temperatures will be experienced through both a gradual change in temperature and in an increase in the frequency and intensity of extreme climatic events (Engelbrecht et al., 2015; IPCC, 2021). For tourists, very hot or cold days during their vacation, particularly when such extreme events are aseasonal, are acutely experienced as uncomfortable weather (Fitchett & Hoogendoorn, 2019; Fitchett et al., 2020). In Europe, extreme temperature events such as the 1998 heatwave in Greece have already been responsible for the death of tourists (Perry, 2006). The more gradual increase in mean temperature is difficult for residents or tourists to detect, and they may only be aware of these changes if they are tracking long-term meteorological records or a critical threshold is exceeded. Indeed, gradual warming is often detected by tourists abroad indirectly through the impacts on glacial retreat (Abrahams et al., 2021). These gradual changes in mean climate have a progressive impact on the climatic suitability of a destination (Fitchett et al., 2017; Mushawemhuka et al., 2020; Noome & Fitchett, 2021) and heighten the probability of critical thresholds being surpassed. It is therefore very important that these changes are carefully monitored, and that their spatio-temporal heterogeneity is understood to develop effective planning strategies for the tourism sector.

The southern African snow tourism sector is experiencing the most urgent threat of temperature and rainfall thresholds being exceeded (Hoogendoorn et al., 2021). Afriski and Tiffendell Ski Resorts are located within the Drakensberg-Maloti Mountains of southern Africa, a region which currently experiences an average of 11.5 snowfall events per annum (Grab et al., 2017). While this is significantly more snowfall than anywhere else in southern Africa, it is insufficient to ensure a reliable ski season. Afriski lodge therefore heavily augments this with artificial snow production using snow guns (Hoogendoorn et al., 2021), operated through careful real-time monitoring of meteorological conditions at the top and bottom of the ski slope. Under continued temperature increases and changes to humidity levels and precipitation rates, more extensive and expensive approaches such as snow factories may become necessary, approaches already under consideration by Afriski lodge management (Hoogendoorn et al., 2021). A reduction in the frequency of heavy snowfalls and days of frost will negatively affect the destination image of these ski resorts, as many tourists are very keen to experience snow first-hand (Stockigt et al., 2018). The temperature increases are also of concern to the comfort of the tourists while skiing, with several ski-tourists at Afriski already indicating that they need to strip down to cooler clothing during the warmest hours of the day (Stockigt et al., 2018).

Gradual changes in mean climate are also influencing the tourism sector indirectly through resultant changes in the timing of annually recurrent biological events, such as flowering, which may form the primary attraction of some destinations. These are termed phenological shifts, driven by even slight changes in temperature and rainfall (Fitchett et al., 2015). One example is the famous annual Namaqualand Daisy flowering season along the west coast of South Africa each spring. Tourism in the region is dependent on tourists visiting the Namaqualand during flowering season, with tour operators, accommodation establishments, and restaurants receiving a necessary cash inflow with an influx of tourists for the few weeks of the flowering season. However, as a result of gradual changes in temperature and rainfall in the Namaqualand region, the flowering dates are advancing at a rate of 2.1–2.6 days per decade and heightened inter-annual variability in flowering dates (Snyman, 2020). These changes in the timing of flowering, if not closely monitored and communicated throughout the tourism sector, could result in tourists arriving after peak flowering had occurred, or missing the flowering altogether, resulting in disappointed tourists, contracted visits, and potential cancellations (Cowling et al., 1999; Preston-Whyte & Watson, 2005; Snyman, 2020). For wedding tourism, destinations and venues rely on floral arrangements and decorations (Rogerson & Wolfaardt, 2015), and for some venues the floral setting, such as cherry orchards in blossom, forms the primary reason for the choice of location and wedding date (Mahlangu & Fitchett, 2019). The increasing variability in flowering dates that is occurring due to changes in late winter temperatures and the timing of onset of spring rains results in an increasing chance that the flowering season may not coincide with wedding dates that were booked months to years in advance, leading to customer dissatisfaction and the potential for a reduction in future bookings of the venue (Mahlangu & Fitchett, 2019).

BOX 5.1 ARID REGION TOURISM IN SOUTHERN AFRICA

Kirsten Noome, Ian Steyn, and Kayla Mac Conachie

Semi-arid and arid regions in southern Africa (most notably Namibia) are characterised by highly variable seasonal rainfall (inter-annually and intra-seasonally), frequently leading to both droughts and flash floods and temperatures that are projected to increase between 1°C and 4°C by 2050 (Spear et al., 2018). Over the past 40 years, Namibia has experienced recurring droughts, heavy rainfall events, episodes of higher temperature, and unpredictable and variable rainfall. Rainfall is expected to decrease by 4% and 7% when global averages reach 1.5°C and 2°C warming, respectively (Hoegh-Guldberg et al., 2018).

Water is a fundamental resource for arid regions such as Namibia and is extremely limited (Spear et al., 2018). Changes in fog patterns, possibly caused by El Niño events or global climate change, could have major implications on the availability of this important source of water in the desert. Fog is necessary to the survival of these desert species, but to what degree is still unknown (Mitchell et al., 2020). Modelling the impact that climate change will have on fog intensity, frequency, and duration is challenging (Vale & Brito, 2015). Some models suggest that the increase in aridification will cause the fog-belt to become smaller, affecting hundreds of desert species (Hänsler et al., 2011). The expected increase in weather-related disaster risk poses significant challenges to tourism sectors in Namibia (Tervo-Kankare et al., 2018; Wilhelm, 2013). Nature and cultural tourism in Namibia relies solely on its natural resources base, and any impacts to biodiversity and natural ecosystems, including water pollution, land degradation, soil erosion, discharges into the sea, loss of natural habitat for flora and fauna, or ecological disruption, will have a direct effect on tourism (Christie et al., 2013; Eckert, 2020).

Increasing ambient temperatures and unpredictable rainfall resulting from climate change will decrease the survival probability of wildlife species, and sparse plant availability for herbivores will severely impact the ecosystem on multiple levels (Lehmann et al., 2020). A loss of biodiversity or disruption to the local ecosystem has significant effects on tourism to that region, for example, food resource fluctuations in Namibian deserts are driving non-migratory ungulate movements (such as gemsbok), which may impact local tourism safari game drives and sightings (Lehmann et al., 2020).

Rising temperatures have led to a gradual shift in the tourist season around the world. Higher maximum temperatures pose a threat to outdoor tourism through the potential of heat stress to the tourists, causing the outdoor activities to become uncomfortable (Mushawemhuka et al., 2020). According to the TCI calculated for Namibia, summer tourism is unfavourable in terms of thermal comfort compared to winter and autumn seasons for tourists (Fitchett & Noome, 2021). Rising temperatures (and a high rate of evaporation associated with desert climates and limited water resources) may also contribute to high water consumption as well as affect the quality of the water or increased transmission rates of water-borne diseases in camping grounds across Namibia (Baker & Mearns, 2017). Quality drinking water is essential for guest satisfaction, and more importantly, for guest and staff well-being (Baker & Mearns, 2017). In addition, rising temperatures are likely to cause other impacts to tourist well-being, including heat-related stress, such as a stroke, increases in a malaria-carrying mosquito as well as tick population are likely to become increasingly widespread (Garland et al., 2015).

The concept of an integrated, sustainable development strategy for tourism is still in its infancy for many arid regions in Namibia and they are often plagued by water stress which is a limiting factor for economic growth and development. Without effective strategies to manage the threats of climate change on outdoor tourism, tourist satisfaction will decrease, and the probability of repeat visitation will be unlikely (Hoogendoorn & Fitchett, 2018a; Pandy & Rogerson, 2018). Thus, tourism providers should offer alternative tourism products during the off-peak seasons, seasons that may deter tourism due to heat stress or other risks to tourism comfort and satisfaction, and take advantage of the climate. Where temperatures or fog cover may disrupt the tourist experience, the tourists' journeys *en route* between destinations, including organised photographic, sightseeing and hunting safaris, self-drive safaris, birding, off-road 4×4 routes, nature trails, cultural activities, historical sites and folklore, and more, will increase the tourist experience and bring awareness to different tourism resources (Fitchett & Noome, 2021).

These phenological shifts, and their impacts of on tourism, are not limited to plants. The 'sardine run', the annual migration of sardine from the cold water Agulhas Bank to the warm Indian Ocean off Mozambique during the austral winter, attracts angling tourists to the east coast of South Africa and is important to fisheries. The timing of the sardine run has become increasingly unpredictable and progressively been delayed by 1.3 days per decade since the 1940s (Fitchett et al., 2019). These progressive delays influencing the timing and quality of fish migrations, coupled with declining fish stocks due to overfishing, have long-term detrimental impacts on angling tourism in southern Africa at large (Hoogendoorn, 2014). For mammals, the progressive changes in climate are expressed through changes in the timing and range of migration. This is particularly detrimental where national parks inhibit natural migration paths (Thuiller et al., 2006). Under climate change, migration is progressively displaced latitudinally and altitudinally (IPCC, 2021). Where park boundaries prevent these changes to migration, animals will be unable to move to sufficiently climatically different settings and thus experience heat stress and dehydration (Hoogendoorn et al., 2019; Preston-Whyte & Watson, 2005). Transfrontier Parks and parks with unfenced borders address these challenges to an extent (Dube et al., 2020; Hoogendoorn et al., 2019), but cannot address the problem indefinitely (Hoogendoorn & Fitchett, 2020). At much shorter time periods, seasonal changes in migration patterns as temperature and rainfall patterns begin to change have an impact on the predictability of animal sightings at particular locations such as watering holes (Smith & Fitchett, 2020).

Given the spatial heterogeneity in both the mean contemporary climate and climate change projections, the close monitoring of climatic trends to detect these gradual changes needs to be performed at high spatio-temporal resolution (van der Walt & Fitchett, 2021c). While a couple of studies on tourism and climate change in southern Africa have presented trend analyses of meteorological data of a location of interest (see Dube & Nhamo, 2020b), or climate indices such as the Standardised Precipitation Index (Smith & Fitchett, 2020), it could be argued that these lack context without incorporating thresholds for tourist comfort, satisfaction, and enjoyment. It is for this reason that a range of tourism climate indices have been developed, and increasingly, longitudinal studies of these index scores are being published for southern Africa (Fitchett et al., 2017; Mushawemhuka et al., 2020; Noome & Fitchett, 2019), quantifying these gradual changes in climatic suitability for tourism.

Threats of drought

Drought, scientifically defined as a prolonged significant departure from mean annual rainfall, has presented the most frequent and extensive climate-related losses to the southern African tourism sector thus far (see Dube et al., 2020; Smith & Fitchett, 2020). As a semi-arid region, southern Africa faces chronic water shortage (Tervo-Kankare et al., 2018), which during drought events can result in stringent water restrictions, the closure of leisure facilities such as swimming pools, and the suspension of water-dependant outdoor activities such as white-water rafting (Dube et al., 2020; Giddy et al., 2017a; Prinsloo, 2019). Drought events also have a marked impact on the natural landscape, affecting the vegetation cover and density, altering animal migration patterns, and in the most severe events, resulting in the death of fauna and flora (Smith & Fitchett, 2020). While climate events such as tropical cyclones and tornadoes result in costly destruction of infrastructure, and extreme heat events threaten the health and comfort of tourists, drought events are much longer lasting (months through years), affect large geographical regions, and therefore arguably have a longer-lasting impact on destination image (Tegegne et al., 2018).

BOX 5.2 DROUGHT   IMPACTS   ON   NATURE-BASED TOURISM

Tamzyn Smith

Droughts – periods with significantly lower precipitation levels than what is considered normal for the spatio-temporal context (Svoboda et al., 2012) – are posited to increase in frequency, severity, spatial extent, and duration under climate change. NBT is driven primarily by the flora, fauna, and aesthetic appeal of a location (Kutzner,

2019) – natural capital elements which are underpinned by climate variables including rainfall (Mushawemhuka et al., 2018). NBT is therefore particularly vulnerable to the impacts of drought (Smith & Fitchett, 2020). Although there are limited reports of drought affecting the number of visitors travelling to NBT destinations (Mathiva et al., 2017; Smith & Fitchett, 2020), there have been multiple instances of droughts negatively impacting the experiences of tourists and the broader functioning and profitability of the sector (Dube et al., 2020; Smith & Fitchett, 2020).

Dry riverbeds, dams, and desiccated vegetation caused by droughts impact the aesthetic appeal of a location (Smith & Fitchett, 2020). This is further damaged by extreme biomass loss (Abraham et al., 2019), as tourists have expectations informed by their imagined geographies but are met with what they perceive to be degraded environments, thus negatively altering the tourist destination image (Prinsloo, 2019; Smith & Fitchett, 2020). Biomass loss also limits food supply for herbivores, subsequently reducing populations (Abraham et al., 2019). Although it can be argued that these impacts make future generations more resilient to droughts through natural selection, it nonetheless negatively impacts the game viewing experiences of tourists (Serdeczny et al., 2016; Smith & Fitchett, 2020). High biodiversity levels are associated with an increase in NBT (Chung et al., 2018). However, biodiversity decreases during drought periods (Serdeczny et al., 2016), further damaging tourist game viewing experiences. The hydrological impact of drought reduces the amount of water that is available for use from rivers, dams, and boreholes, limiting recreation activities, water for baths, swimming pools, and regular cleaning of game viewing vehicles.

NBT suffers both financial and operational impacts due to drought (Smith & Fitchett, 2020). These impacts are difficult to quantify, often arising due to increased running costs relating to higher food prices and maintenance costs including water pipes dug up by elephants, fences broken by animals expanding their range in search of water, and water pumps requiring repair following exposure to air (Smith & Fitchett, 2020; Steyn & Spencer, 2012). Additional costs relating to drought include veterinary callouts and ecological interventions such as animal relocation (Smith & Fitchett, 2020).

The TCI suggests that a reduction in rainfall improves the climate suitability of a destination (Fitchett et al. 2017), but in the extreme case of drought, this is not necessarily the case. Drought events can have complex and detrimental impacts on NBT that can extend long after drought conditions have ended (Swemmer et al., 2018) and can be spatiotemporally heterogeneous (Smith & Fitchett, 2020). Acknowledging that NBT is particularly vulnerable to climate change (Mathiva et al., 2017) and understanding the impacts of drought is critical for adaptation.

The most recent and best-known example of a southern African drought that had major impacts on tourism was the 'Day Zero' drought in the City of Cape Town (see Chapter 3) that spanned 2015–2018 (Dube et al., 2020; Wolski, 2018). Prior to the 'Day Zero' drought, the impacts of drought in southern Africa were most acutely experienced in the NBT sector, and as with many climate threats to tourism, urban tourism was largely considered to be immune (Fitchett, 2021). The effects of drought are still common in the NBT sector, affecting both tourist comfort through similar water restrictions on bathing, showering, and swimming and tourist enjoyment through changes to the natural environment. For the Sabi Sands Game Reserve in South Africa, the impact of the 2015 drought on tourist numbers was limited due to the operator's awareness of the climate threats and proactive measures (Smith & Fitchett, 2020). The impacts on the animals, however, were marked, requiring the costly relocation of large animals to prevent further mass mortality (Smith & Fitchett, 2020). For the Kruger National Park, Dube and Nhamo (2020a) noted that long-term droughts have decreased the fuel load of grass for fires and consequently lead to exacerbated bush encroachment. Natural watering holes for animals invariably dry up during drought periods, which leads to water stress and eventual death for animals that frequent them, including elephant, rhinoceros, and buffalo, and reduce visitor's opportunities to view animals at these popular spots (Smith & Fitchett, 2020).

Within the adventure tourism sector, attractions that are reliant on rivers, dams, and lakes are particularly vulnerable to droughts (Buckley, 2017). White-water rafting, for example, relies on sufficient river levels to allow for safe activities and, in turn, attract sufficient income (Carolli et al., 2017). In the case of South African and Namibian white-water rafting, drought was found to be the second most damaging occurrence for white-water tourism, ultimately leading to increased trip cancellations and closing of business in extreme cases (Giddy et al., 2017a). Mushawemhuka (2021) similarly found that low water levels induced by drought inhibit business operations on the Zambezi River including white-water rafting, with a more sensitive water-level threshold for activities involving larger boats. Drought is also a concern for snow tourism in Lesotho as well-supplied water storage dams are required for the production of artificial snow and winter precipitation for natural snowfalls (Hoogendoorn et al., 2021). Insufficient water levels also have an impact on angling tourism events such as the Wild Trout Association Fly Fishing Festival held annually at the village of Rhodes in the North-Eastern Cape of South Africa. During drier conditions in 2013, anglers at this festival only caught 93 trout, while during the wetter year of 2016, anglers caught more than 1700 trout (Hoogendoorn & Fitchett, 2018b). For Lake Kariba, drought threatens both the local biodiversity and water levels of the lake, affecting both nature-based and adventure tourism (Dube & Nhamo, 2020b; Mushawemhuka, 2021). These impacts have resulted in a reduction in tourists' arrivals, lowering occupancy rates in hotels and threatening the viability of tourism business in adjacent towns (Dube & Nhamo, 2020b).

While arid regions are arguably better adapted to dry conditions, Tervo-Kankare et al. (2018) note that Namibia, an arid to hyper-arid country (Göttsche & Hulley, 2012), is more prone to drought than other regions of southern Africa. The regular exposure to drought conditions has widespread consequences. For example, in the Namibian village of Uis, drought has impacted visitation numbers and adversely affected the income of tourism businesses (Tervo-Kankare et al., 2018). Similarly, tour operators in the Kalahari Desert of Botswana are concerned about future livelihoods under climate change and the socio-economic impacts of drought for the region (Saarinen et al., 2012). The same concern applies to the broader socio-economic consequences for local communities that are dependent on tourism should droughts increase in frequency or severity (see Kaján & Saarinen, 2013; Saarinen et al., 2020).

Victoria Falls has been a key site of climate change and tourism research in southern Africa, particularly in relation to drought (Dube & Nhamo, 2019a; Mushawemhuka et al., 2021). There was public outcry during late 2019, especially on social media, regarding the reduced flow of Victoria Falls and the Zambezi river, with claims that the waterfall is drying up (Dube & Nhamo, 2020c; Mushawemhuka et al., 2021). However, the veracity of these claims, and the role of climate variability versus climate change, is contested (Mushawemhuka et al., 2021), and incorrect information in this regard may be more detrimental to the tourism sector than the actual reduction in water flow. NBT operations have been negatively affected by the perceived drought in the region through both impacts on activities that tourists can take part in and the damage to destination image (Dube & Nhamo, 2020c; Mushawemhuka, 2021; Mushawemhuka et al., 2021).

Critical to both effective planning for extreme climate events and preventing harm to tourist visitation numbers through misinformation is the accurate classification of drought events and effective modelling of future drought probabilities (Smith & Fitchett, 2020). The southern African region is characterised by considerable inter-annual variability in mean annual precipitation, resulting in few statistically significant trends in historical rainfall amount and considerable spatial heterogeneity in future precipitation projections (Kusangaya et al., 2014). Much of southern Africa is projected to become drier in future decades (IPCC, 2021). Drought intensity is projected to increase across the region, particularly when classified by evapotranspiration (Abiodun et al., 2019; IPCC, 2021). Despite the threats of drought for tourism, much of the research on drought in the tourism context fails to classify droughts according the World Meteorological Organization Standards (Smith & Fitchett, 2020), which has resulted in the conflation of drought and dry periods, and drought and aridity, by tourism researchers, operators, and tourists alike. Adding to this challenge, tourism-specific climate indices such as the TCI classify low rainfall as preferable and therefore misattribute drought events as being favourable climate conditions for tourists (Noome & Fitchett, 2019). Greater focus on appropriate metrics in drought-prone settings and more accurate use of terminology is therefore important in preparing for a future of heightened rainfall variability and changes in drought frequency (Kusangaya et al., 2014).

Threats of severe storm events

Whereas mean annual temperature increases demonstrate an approximately linear trend under climate change, precipitation changes are spatially and temporally heterogenous (Engelbrecht et al., 2011; IPCC, 2021). Therefore, both arid and more humid regions, those which experience frequent droughts and those which do not and those with projected rainfall increases and those of rainfall decreases, remain at risk of severe storm events (IPCC, 2021; Mason & Joubert, 1997). For southern Africa, these extreme storm events are predominantly driven by tropical storms and cyclones, and cut-off lows, as described in Chapter 2. These storm events are associated with heavy rainfall, which can induce flooding, lightning, hail, and strong winds, with the potential for storm surge along coastal regions. The majority of these extreme storm events persist for only a few hours, but result in damage to infrastructure that can take months to years to repair (Fitchett et al., 2016c). Extreme storm events are projected to increase in both intensity and frequency of occurrence under climate change (IPCC, 2021).

Tropical depressions, storms, and cyclones form between 5°S and 25°S in the southwest Indian Ocean, making landfall on Mauritius, Reunion, Madagascar, Mozambique, and the Comoros, and with storm tracks extending into South Africa, Zimbabwe, and Botswana (Pillay & Fitchett, 2021). Mahadew and Appadoo (2019) argue that the frequency of extreme storm events and heavy rains associated with tropical cyclones in Mauritius has increased over the past two decades. Mauritius is the most tourism-reliant country within in southern Africa, and tourism is the third most important economic sector for Mauritius (Fitchett et al., 2020). Therefore, changes in storm frequency pose a serious threat to economic survival of both Mauritius and its tourism sector. An analysis of 13,618 TripAdvisor reviews of Indian Ocean Islands spanning the period 2012–2016 confirmed that tourists to Mauritius were particularly aware of the risks of tropical cyclones (Fitchett et al., 2020). However, the threats are not unique to Mauritius. Across the South Indian Ocean, the number of tropical cyclones has decreased in recent decades, while the intensity of the storms has increased with the emergence of category 5 tropical cyclones in the past three decades (Fitchett, 2018; Fitchett & Grab, 2014). There has also been a poleward expansion in the geographical range of storm tracks, extending outside of the tropical region to 30°S, heightening the storm risk for southern Madagascar, southern Mozambique, and north-eastern South Africa (Pillay & Fitchett, 2019).

Lower intensity tropical storms are more common, but cause considerable damage to tourism-dependent towns and tourist operations. One documented example is of Tropical Storm Dando which made landfall in January 2012, affecting the Mopani District Municipality of South Africa (Fitchett et al., 2016c). This storm caused major flooding in the low-lying regions of the Limpopo province, with over 500mm of rain over a three-day period (Fitchett et al., 2021), resulting in major infrastructural damage to roads, water provision, accommodation establishments, and services in the area (Fitchett et al, 2016c). The higher-intensity Tropical Cyclone Dineo, which made landfall on the coast of Mozambique in February 2017, notably is recorded to have resulted in less damage than Tropical Storm Dando (Fitchett et al., 2021). Dineo resulted in uprooted trees falling on accommodation, and damage to orchards, and some water damage to accommodation, broader infrastructural damage of the local towns, and municipalities such as damage to roads, bridges, and electricity blackouts (Fitchett et al., 2021). The lower extent of damage is likely due to the relative paths of the two storms.

BOX 5.3 CLIMATE CHANGE THREATS TO HERITAGE TOURISM IN SOUTHERN AFRICA

Fazlin McPherson

There is growing research interest on both climate change and heritage tourism, but very little specifically focuses on the relationship between heritage tourism and climate change (Hall, 2016). South Africa has several natural and cultural heritage sites spread across the country, including Robben Island, The Cradle of Humankind, Drakensberg, and the Cape Floristic Region. These sites are internationally recognised and attract local and international tourists annually (Rogerson & van der Merwe, 2016). Heritage tourism is quickly becoming part of a huge global sector of cultural tourism (Kourtit et al., 2019). It has been suggested that heritage tourism should be seen as one of the important contributors to the South African economy (Khumalo et al., 2014). This is because the country hosts some of the most prominent rock art and archaeological sites at Mapungubwe, Mokopane, The Cradle of Humankind, Makgabeng, and the West Coast Fossil Park, to name a few. Heritage sites can include both cultural and natural sites which, in turn, form part of the environment and are therefore vulnerable to climate change (Tompkins & Adger, 2004).

Makgabeng is one of the many tourism destinations in southern Africa that offers cultural heritage tourism and natural heritage tourism.

Makgabeng is situated in the northwest corner of the Limpopo Province in South Africa. The Makgabeng and the Blouberg mountains are of extraordinary, picturesque beauty, with a mixture of a bushveld-savanna bio and yellow-wood forests (Van Schalkwyk, 2009). Some of the tourist offerings within the area include hiking, mountain climbing, and hunting. Some of the other activities include features such as access to some of the most important archaeological sites. The Makgabeng is best known for the abundance of rock art, with just over 600 rock art sites. The communities who are presently living in the area are a heterogeneous accumulation of people who had come to the area through a complex set of processes during the Apartheid era (Van Schalkwyk, 2009). As a result, the area hosts a unique living heritage, including architecture, food, dances, digging sticks, mats, wooden spoons, traditional beer, clay pots, basketry, poetry, music, and traditional costumes.

In southern Africa, many heritage tourism attractions are outdoors, so rain and extreme temperatures can pose challenges. In a region such as the Makgabeng where the primary attraction is heritage tourism, climate change poses a severe threat (McPherson, 2020). For outdoor tourism, especially the rock art sites, hot conditions during summer months limit the amount of time that tourists can spend at the attractions, and the heat and sun makes hikes to rock art sites strenuous (McPherson, 2020). Heat is also a problem for indoor cultural and heritage tourism attractions, requiring air-conditioning for both human comfort and for the preservation of artefacts. Odeku (2008) suggests that there is a deep concern around heritage sites being vulnerable to various threats such as the impact and effect of global climate changes, mining pollution, and other hazardous human activities. This may result in the degradation of these heritage sites and will consequently lower their tourism value and importance (Odeku, 2008). Climate change could destroy certain archaeological artefacts as well as archaeological remains. It could also destroy the natural environment of a destination by destroying flora and fauna through heavy rains or extreme weather conditions. Many of these tourism destinations hold a dual attraction of both natural and cultural heritage such as game viewing and hiking. Therefore, climate change will ultimately lower the value of these heritage sites and ability to attract a wider range of tourists. It is of utmost importance that heritage tourism operators across southern Africa carefully consider the climate change threats to their attractions and develop effective adaptation strategies.

Two severe tropical cyclones have made landfall on the southern African subcontinent in recent years – Cycones Idai and Eloise, both of which made landfall on Beira, Mozambique. The impacts of the two storms are staggering, causing a humanitarian disaster (Devi, 2019). Research on the impacts of Cyclone Idai on tourism is slowly emerging (see Sepula & Korir, 2019), and investigations into the impacts of both storms are no doubt still underway. It is notable, however, that very little has been written on tropical cyclones and tourism in the region of frequent storm tracks. In broader terms, flooding resulting from extreme storm events has been experienced in the white-water rafting (Giddy et al., 2017a) and NBT sectors (Mushawemhuka et al., 2018; Southon & van der Merwe, 2018) of southern Africa. Further research into extreme storm events associated with a diverse set of climatological conditions and of the impacts of hail, lightning, and storm surge, and what the IPCC (2021) terms compound extreme events are important in understanding the nature and extent of the impacts on the tourism sector.

Threats of extreme temperature events

In addition to gradual increases in temperature, which will eventually exceed human comfort thresholds, increases in both extreme heat and cold temperature events have been recorded in southern Africa over recent decades (New et al., 2006; van der Walt & Fitchett 2021a, 2021b) and increases in extreme heat events are projected to increase in frequency and severity (IPCC, 2021; Mbokodo et al., 2020). Extreme heat events pose an acute health hazard to tourists and local tourism operators alike, with severe events leading to heat stroke, cardiovascular issues, and the potential for death (Wright et al., 2021). The increase in the incidence of heatwaves in southern Africa is likely to impact the tourism sector at large, forcing tourists to stay indoors for large parts of their holiday (van der Walt & Fitchett, 2020a). It is notable, in this context, that in surveying American tourists visiting South Africa, the majority of respondents indicated that they did not feel too hot or too cold during their visit (Giddy et al., 2017b). Similarly, when surveying beach tourists along the south and east coasts of South Africa, 13% of respondents perceived cold conditions to be a cause for trip cancellation, while only 3% would terminate a trip due to hot conditions (Friedrich et al., 2020b). Extreme temperatures are more frequently raised as a point for concern among tourists and tourism operators in regions of Zimbabwe (Mushawemhuka, 2021) and Botswana (Saarinen et al., 2012).

In the case of the airline industry, Dube and Nhamo (2019a, 2019b) argue that greater levels of air turbulence occur when temperatures exceed 35°C, necessitating longer runways and load-shedding by airplanes. This would detriment airplane performance levels, which would have an impact on ticket pricing and insurance (Dube & Nhamo, 2019b). Dube and Nhamo

(2019a) present the example of an airplane that could not land at the Kariba Airport (Zimbabwe) in November 2015 when temperatures exceeded 40°C, due to concerns of tyres bursting or engine damage. The airplane was consequently rerouted to Victoria Falls International Airport (News 24, 2015). It is important to note that this was an isolated incident, and that empirical studies would be needed to determine whether concerns of repeated events, or increases in ticket price and insurance costs are valid. At present, this seems unlikely, given that the United Arab Emirates run major international airports with considerably higher levels of air traffic, under climatic conditions of much warmer mean temperatures than southern Africa. As most flights to Kariba Airport and Victoria Falls International Airport originate from OR Tambo International Airport in Johannesburg, the short flight of 100 minutes is unlikely to be frequently affected.

The urban heat island heightens the frequency and severity of extreme temperature events, raising the threats to urban tourism (Fitchett, 2021; van der Walt & Fitchett, 2021a, 2021b). Urban tourism is the largest form of tourism in southern Africa, driven particularly by retail tourism (Rogerson, 2011). Dominated by regional tourists visiting major metropoles such as Johannesburg or Durban, these cities host millions of tourists from countries bordering South Africa (Rogerson, 2011). Movements to and from these countries are especially important in terms of the economic survival of both the source and host nations. While urban tourism is often considered to be less sensitive to the climate and weather, as much of the attractions are indoors in climate-controlled settings (Fitchett, 2021), air-conditioning is not as ubiquitous in southern Africa as in much of Europe or North America. Outdoor events such as urban walking tours in major cities are particularly at risk, especially when walking between attractions over multiple hours (Hoogendoorn & Giddy, 2017).

Threats of changing wind regimes

Depending on the speed and consistency, and the attraction of interest, wind can be either an asset or a hazard to tourism. Strong and consistent winds are important for wind surfing destinations, while more gentle breezes are preferred among beach tourists (Becken, 2010). For adventure tourism, strong winds can be dangerous, interrupting events until the winds subside (de Freitas, 2003). Changes in wind patterns can therefore make a region either better or worse suited to a given activity, dependant on the optimal threshold and whether it has been exceeded.

The impact of changing wind regimes is a very under-studied component of tourism and climate change research in southern Africa. TripAdvisor reviews provide a broad sense of tourists' experiences of strong winds, whether anomalous or typical for a region (Fitchett & Hoogendoorn, 2018; Fitchett et al., 2020). For the Comoros and Reunion, wind was jointly the most cited climatic concern of tourists in their reviews, while for Mauritius, it was the second most frequently mentioned (Fitchett et al., 2020). For some reviewers, the strong prevailing winds prevented them from taking part in outdoor activities, leading to the early termination of their trip (Fitchett et al., 2020). For South Africa, by contrast, wind only accounts for 2.5% of reviews mentioning climate, although with a degree of variability between destinations (Fitchett & Hoogendoorn, 2019). It is notable that when interviewing American tourists to South Africa, wind was the most frequently cited weather phenomena to hamper the tourist experience (Giddy et al., 2017b). The discrepancy may be due to the destinations tourists and TripAdvisor reviewers had visited, with a dominance among American tourists for the coastal destinations of Gqeberha, the 'Windy City' of South Africa, and Cape Town, which is renowned for excessive winds such as south-easter called the 'Cape Doctor' (Giddy et al., 2017b; Kruger et al., 2010).

In South Africa, the effects of wind are more short-lived than for the Indian Ocean Islands, but when ill-timed can have significant consequences to tourist events. The Cape Town Cycle Tour (CTCT) was cancelled in 2017 due to extreme winds that would have posed significant danger to cyclists and spectators (Giddy, 2019). The CTCT had become known as the 'Tour de Storms', with windspeeds of up to 120 km/h recorded during some years (Giddy, 2019, p. 99). In the survey conducted by Giddy (2019), most respondents noted that wind was the primary concern for not taking part in this event.

A particular concern related wind in southern Africa is the threat of ignition and spread of fire and the resultant impact on tourist destinations. The management of controlled burning is not a new phenomenon, and NBT destinations in southern Africa have relatively effective fire management strategies in place (Spenceley, 2005). However, with the increase in the severity and regularity of fires, induced by both climate changes and anthropogenic influence on landscapes, fires are becoming more unpredictable and what the IPCC (2021) terms fire weather has become more common (Archibald, 2016; Smit et al., 2013). Hot, dry katabatic winds, locally termed Berg winds, which form coastward of the escarpment particularly in winter, further heighten the risk of fire and the consequent fires (Kraaij et al., 2013). Fire is therefore of concern to several tourist destinations along the southern coast of South Africa. The most severe example to date was Knysna fire of June 2017, which resulted in major infrastructural damage around the major tourist town of Knysna (Hoogendoorn, 2021), including the loss of 800 buildings, 5,000 hectares of forest plantation, and the loss of seven lives (Kraaij et al., 2018). The economic fallout from this disaster is still acutely felt in the region, and a variety of management initiatives have been implemented to avoid a similar disaster in future. Second home tourism is particularly at threat, as properties are left unattended, in some cases seasonally, allowing grass and trees to grow excessively during the rainy season, and become dry and ignite during the dry months (Hoogendoorn, 2011b). Without consistent upkeep from the second home owners to avoid this situation, this can place second home towns at risk (Hoogendoorn & Fitchett, 2018b).

Sea-level rise

Global SLR is under lowest emissions scenarios and is projected to be in the range of 0.28–0.55m by 2100, and under the highest emissions scenario projected to range from 0.63m to 1.01m (IPCC, 2021). The southern African coastline is roughly 6,000km in length and is generally quite rocky, with limited low-lying areas (Harris et al., 2011). Thus, the impact of SLR is not as pronounced as in regions such as New Orleans, Venice, or Pacific Ocean islands (Nicholls, 2011). The southern African coastline is, however, projected to experience SLR in excess of the global mean (Ragoonaden et al., 2017), with projected hazardous impacts for the low-lying destinations along the South African, Mozambican, and Namibian coastlines and the Indian Ocean Islands, particularly those which are economically dependent on sea, sun, and surf tourism (Friedrich et al., 2020a, 2020b). It is surprising, in this context, that very few studies have explored SLR in the region and specifically, the impacts on tourism.

Among the existing research on SLR impacts on tourism is that of the coastal tourist towns of St. Francis Bay and Cape St. Francis in the Eastern Cape province of South Africa (Fitchett et al., 2016a). These towns are threatened by SLR enveloping beaches and large parts of the developed areas by 2100, with threats to infrastructure projected during high tides and periods of flooding during storm surges (Fitchett et al., 2016a). For Cape Town, Steyn and Spencer (2012) theorise that similar issues may be experienced, and certain areas could experience persistent buffeting from waves. Steyn (2012) argues that estuaries, lagoons, and areas 'reclaimed' from the ocean will be the first to be affected by SLR. As many tourism attractions are positioned along estuaries, this could have negative impacts on future business operations. Dube et al. (2021a) claim that at the majority of Cape Town's so-called 'Blue Flag' beaches (a criteria for top quality tourist beaches in South Africa) will be eventually threatened by SLR and consequent coastal erosion. They also argue that some of Cape Town's most well-known destinations such as the V&A Waterfront (retail tourism), Robben Island (heritage tourism), Cape Point (NBT), and False Bay (beach and NBT) as well as municipal infrastructure including coastal roads are under severe threat due to SLR.

BOX 5.4 COASTAL TOURISM IN SOUTH AFRICA

Jonathan Friedrich and Jannik Stahl

South African coastal tourism presents tourists the opportunity to pursue a wide spectrum of interests and activities. Coastal tourists are able to combine leisure tourism at the beach with other purposes such as visiting nature attractions and cultural heritage sites, which is a major pull factor for international tourists. Thus, it is of great importance to explore the way in which climate change may affect beach tourism in the future.

The core issues for coastal tourism associated with future climate change according to our studies (Friedrich et al., 2020a, 2020b, 2021) are SLR and climate change-induced land-use changes. SLR, in particular, poses a major challenge to coastal tourism as it threatens beach tourism through beach inundation. Case studies on Cape St. Francis, St. Francis Bay (Fitchett et al., 2016a; Hoogendoorn et al., 2016), and Cape Town (Dube et al., 2021a) show that major reductions in beach area will result from SLR. In Cape Town and Cape St. Francis, by the year 2050 a significant number of the beaches are projected to be affected by rising sea levels and inundation (Fitchett et al., 2016a). This will not only affect beaches but also touristic infrastructure, which, in return, may pose threats to the accessibility of destinations in the future. Furthermore, touristic attractions such as Robben Island could significantly be impacted by SLR and reduce the attractiveness of these sites (Dube et al., 2021a). This also applies to climate change-induced land-use changes which could lead to the destruction of natural heritage, coupled with effects in other sectors that may affect both tourism operators and tourists. Other climate change threats to the South African coastline are *inter alia* storm surges introduced by increased tidal activities, more frequent tropical cyclones, coastal erosion, and coral bleaching.

The respondents of our study (Friedrich et al., 2020a, 2020b) indicated that favourable weather conditions such as the absence of high wind speed, rain, and high humidity would lead them to extend their stay. High temperature was indicated to not influence decisions on the extending or shortening a trip. Extreme events, such as floods and fires, were answered by half of the respondents to lead to cancellation of trips. Using hazard-activity pairs, we explored the influence of future precipitation and temperature change (based on downscaled climate models) on beach tourism (Friedrich et al., 2020b). Our results show that destinations along the Garden Route could benefit from decreasing precipitation and rising temperatures as more days per year would follow what is perceived to be favourable weather conditions for beach tourism.

Each beach tourism destination along the coastline faces a unique set of threats and opportunities under a changing climate as climate change follows a locational heterogeneity. For example, destinations along the north-eastern coastline, such as Durban and St. Lucia, are expected to face increased risk of tropical storms and SLR, while destinations along the Garden Route may rather face threats associated with loss of flora and water scarcity. Tourists visiting these various destinations appear to differ in their perceptions of climate. For example, respondents visiting Durban (an already humid climate) appeared to be less concerned about humidity than tourists visiting other destinations. This also means that each destination needs individual strategies to design adaptive tourism practices under climate change.

Climate and climate change perceptions of beach tourists are correlated with their country of origin. International tourists are less concerned about climatic factors influencing their travel decisions, while domestic tourists appeared to be more sensitive of trip decisions. This means that climate change in terms of changing temperature, precipitation, wind, and humidity could rather have an influence on domestic travel behaviour. However, climate change can still have an effect on international tourist arrivals as the international tourism market could be influenced through international policies and governance mechanisms such as carbon taxation on a global and local scale. This underlines the urgency to develop an adaptive and sustainable tourism sector in South Africa to sustain the economic contribution of the coastal tourism sector.

South Africa's prime beach tourism destination and major metropole, Durban (Ahmed et al., 2008), is projected to be severely affected by SLR. Mather (2007) notes that sandy beaches of the KwaZulu-Natal Province coastline, and Durban's beaches in particular, are a key tourism product for South African domestic tourists – 73% of the tourists visiting the coastal region of the province will visit a beach during their vacation. In addition to the beaches, Mather (2007) argues that if no adaptation measure is put in place, the key tourism product of Durban, Ushaka Marine World, and surrounding infrastructure (Ezeuduji & Nkosi, 2017; Nxumalo, 2017) could be lost. Mather and Stretch (2012, p. 251) advocate for the erection of coastal defences, along the 'Golden Mile' of Durban, which would severely impact the destination image of this well-known tourism attraction. It is promising, however, that Durban's eThekwini municipality has among the most advanced climate change adaptation strategies of the southern African region, as detailed in Chapter 7.

Small Island States are arguably at the greatest risk of SLR (Betzold & Mohamed, 2017). In the southwest Indian Ocean, the low-lying islands of Mauritius, Comoros, and Reunion are particularly low-lying and at risk of inundation (Palanisamy et al., 2014). For the islands of Mauritius and Rodrigues, SLR has accelerated (Ragoonaden et al., 2017), heightening the risk

for coastal roads and beachfront hotels (Ragoonaden, 1997). For the Comoros, the importance of accurate climate change knowledge in leveraging
funding towards adaptation is found critical, particularly given the short
timeframes before SLR has a significant impact on the island (Betzold &
Mohamed, 2017). These lessons apply to each of the southwest Indian Ocean
Islands, but also to the response to SLR and climate change threats across
the broader southern African region.

Conclusion

Climate change poses a range of direct and indirect threats to tourism attractions, accommodation establishments, and travel infrastructure both
globally and in southern Africa. While the most widespread impacts on
tourism to date have resulted from drought events, the sector faces serious threats from storm and SLR-induced flooding, storm damage through
winds, hail and lightning, extreme temperature events, and extreme winds
and associated fire. The distribution of these threats is spatially and temporally heterogenous, with a significant influence of seasonality. Critical to the
survival of tourism operations is an accurate understanding of the nature
of these climate risks and an appreciation of both climate change timelines
and the practical implication of the probabilities of low-frequency but high-
intensity events.

6 Destination resilience, vulnerabilities, and climate change threats

Introduction

Climate change represents a disturbance and instability for various socio-ecological systems. As demonstrated many times in this book, climate change is a major force changing and disturbing the tourism system. Recently, the idea of resilience has emerged as "one of the major conceptual tools [...] to deal with change" (Berkes & Ross, 2013, p. 6), which has been applied in tourism development and destination governance (see Cochrane, 2010; Hall et al., 2018; Lew, 2014; Lew & Cheer, 2018; Saarinen & Gill, 2019). This is not surprising as tourism is known to be vulnerable and highly influenced by changes and conflicts in its operational environments at different scales. This requires new conceptual and practical approaches in the change management for tourism, which is the case especially in the southern African tourism economy (Hoogendoorn, 2021; Hoogendoorn & Fitchett, 2018a; Rogerson & Baum, 2021).

Resilience refers to "the long-term capacity of a system to deal with change and continue to develop" (Stockholm Resilience Centre, 2015, p. 3). In tourism studies, resilience research has evolved in two broad and sometimes interwoven directions (Saarinen & Gill, 2019). First, there has been an interest in the impacts of climate change and natural disasters on tourism and the vulnerabilities of the sector or tourism-dependent communities (Biggs et al., 2012; Calgaro et al., 2014; Hall et al., 2016; Slocum & Kline, 2014). Second, there has been a regional economy and development emphasis with a focus on socio-economic restructuring and livelihood diversification, especially in former industrial areas or rural contexts (Cochrane, 2010; Gill & Williams, 2014; Saarinen et al., 2020). Here, we are specifically interested in the former perspective, which has multiple connections to the regional economy and development aspects of tourism. This climate change focus involves resilience to gradual change, which is also typical for many studies on regional economic restructuring and diversification (see Martin et al., 2016; Martin & Sunley, 2014). Compared to the regional economic perspective, however, climate change represents a slow-onset disaster (United Nations Office for Disaster Risk Reduction, 2020), affecting the southern African region at a different pace in time and space. Furthermore, climate change also induces sudden impacts, shocks, and disasters at tourism destinations

DOI: 10.4324/9781003102618-6

in the form of storms, floods, and heatwaves, for example, calling for proactive destination resilience planning and regional governance.

The World Tourism Organisation et al. (2008, p. 38) have identified climate change as "...the greatest challenge to the sustainability of tourism in the 21st century", although as noted by Dogru et al. (2019, p. 293), "there is no empirical evidence showing whether tourism is more or less vulnerable" to climate change, compared to some other economic sectors. Nevertheless, southern African tourism products are highly dependent on climate; however that vulnerability assessment should also involve non-climatic elements. Variables like the socio-economic development rate, political and economic stability, inequality, poverty, and education levels make regional tourism specifically vulnerable to climate change. Indeed, available resources define how countries and regions can respond to change, as previous research has demonstrated that resource-rich communities have greater stability, and thus resiliency, than those without resources (Adger, 2009; Gössling et al., 2012). These non-climatic elements are sometimes overlooked when considering a tourism destination's capacity to adapt to and resist climatic changes (Dogru et al., 2019; Hyman, 2014; Luthe & Wyss, 2014), but they are increasingly recognised in policymaking and -setting priorities for research (Hall et al., 2018). The United Nations Framework Convention on Climate Change (UNFCCC) Paris Agreement (United Nations, 2016), for example, strongly emphasises the urgent focus on societal vulnerability issues. Furthermore, the United Nations (2021) roadmap for COVID-19 recovery, for example, highlights the elements of community resilience and social issues, which are firmly addressed in South Africa's two national priorities in the context of SDGs focusing on (i) reducing inequalities and vulnerabilities and (ii) building a more resilient and sustainable society (South Africa's..., 2019) (Box 6.1). Therefore, vulnerability and resilience thinking need to involve the estimated changes in climatic conditions with the inclusion of socio-economic and political elements.

BOX 6.1 CLIMATE CHANGE AND SUSTAINABLE DEVE-LOPMENT GOALS IN SOUTHERN AFRICA

Julius Atlhopheng

The Southern African region is among the most vulnerable to the effects of climate change. This includes livelihoods, the agricultural systems, the biodiversity and ecosystems, and the climatic changes and climate disasters which pose threats to sustainable development in the region (ASSAR, 2021). The degree of anticipated impacts at 1.5°C–3°C temperature change scenarios requires urgent adaptation strategies. Thus, there is need to tackle the threats of climate change, whilst advancing the SDGs. This requires a well-coordinated framework to ensure that the local, regional, and international commitments under the 2015 Paris Agreement and the 2030 Agenda are achieved (United Nations, 2015a). SDG achievements are challenged by the global capitalist system, where development exploits natural resources in Africa (Ramutsindela & Mickler, 2020). Kasirye et al. (2020) highlight the weak civil society and private sector, challenges with funding SDGs, limited localisation, a paucity of data, and multiple development agendas that the continent pursues.

Southern Africa is well endowed with tourism resources (Christie et al., 2013) from breathtaking landscapes to abundant wildlife, diverse opportunities for tourist activities, cultural tourism, and other avenues. Tourism is seen as an agent of economic diversification and offers the potential for sustainability. Yet, tourism is susceptible to climate change and needs to address market failures which may create pockets of poverty in an oasis of abundance. The land and ecosystem services are also noted for their importance as they keep the planet in a steady state or equilibrium. Land has tipping points, for example, soil quality, pollution levels, and climate change. Businesses are derived from the land; hence need to protect this productive resource (land degradation neutral world – SDG15). Climate change, according to the United Nations High Level Political Forum in 2019 (UNFCCC, 2019), has multiplier threats to many facets of the SDGs (see Figure 6.1.1). The United Nations have ventured on a path of collaboration

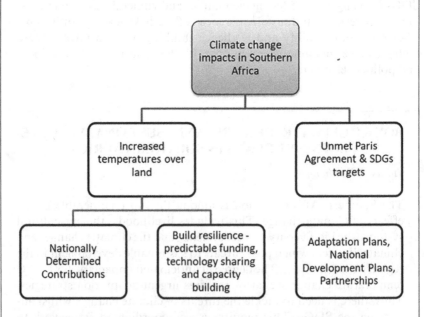

Figure 6.1.1 The dual challenges of climate change and SDGs.
Source: the author.

to overcome climate change, for example, through adaptation, mitigation, and intended national determined contributions.

Interaction of climate change with the SDGs

Sub-Saharan Africa is facing several challenges in attaining the SDGs, including ensuring that no one is left behind. The development trajectories, although well meaning, have resulted in inequalities, for example, a high Gini coefficient demonstrating unequal access to national wealth. The people-planet-peace-prosperity and partnerships (the 5Ps that sum up the SDG aspirations) need improvement, as per the SDGs tracker (UN Statistics Country Profiles, 2018). The climate action and the SDGs outcomes are about making sustainable impact, building synergies between the Paris Agreement and the 2030 global agenda (United Nations, 2015b). These may be in the national development plans, the nationally determined contributions, adaptation strategies, and disasters risk reduction approaches. In fact, New and Bosworth (2018) considering the impending climate change impacts that face Botswana and Namibia under a 2°C scenario have advised on transformations of livelihoods to cope with a warmer regional climate which may be beyond adaptation. The impact of implementing the SDGS, according to Jimenez-Aceituno et al. (2020), may be evident in local level initiatives, such as projects (specific or general), in cross-cutting issues and/or specific SDG implementation. This may be more impactful than the policy approach and may showcase results of high priority SDGs versus those under-represented (in implementation). There is need for data strategies to enhance coordinated implementation. Southern Africa needs urgent action and plans, partnerships, and resources to achieve the dual obligations of the SDGs and climate change (Figure 6.1.2).

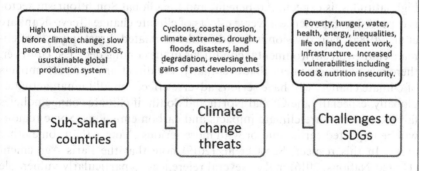

Figure 6.1.2 The climate change-SDGs interactions in southern Africa.
Source: the author.

From vulnerabilities to resilience thinking

Vulnerability

There are different perspectives and ways to understand vulnerability. In the context of climate change, vulnerability is often defined as a degree to which an ecological and/or socio-economic system is susceptible to and unable to cope with the adverse effects of the change (McCarthy et al., 2001). Scott et al. (2019) have further stated that vulnerability is a multi-dimensional concept that cannot be observed or analysed directly. Therefore, it must be constructed by using other measures (Hinkel, 2011). While developing suitable practical measures for vulnerability has been challenging, the idea has been "a powerful analytical tool for describing states of susceptibility to harm, powerlessness, and marginality of both physical and social systems, and for guiding the normative analysis of actions to enhance well-being through reduction of risk" (Adger, 2006, p. 268).

Vulnerability is in a central position in climate change discussions and policies in southern African tourism (Hoogendoorn & Fitchett, 2018a; Hoogendoorn et al., 2021). Rogerson (2016, p. 322) has stated that "South Africa is one of the most vulnerable parts of the world in terms of projected climate change". Considering that South Africa is the most resource-rich country in the region, the wider sub-continent is even more vulnerable to the impacts of climate change. In general, vulnerability to climate change is expected to have a negative impact on economies, and thus, regional tourism sectors and related socio-economic activities are expected to suffer from the impacts of climate change (Scott et al., 2019). The main vulnerability and a direct threat to the southern African tourism sector will be the inability to attract visitors (see Lew & Cheer, 2018) in changing conditions.

Climate change itself, however, may create both winners and losers in the short term. In this respect, some forms of tourism may gain economically while some others lose in southern Africa. Cultural tourism and shopping tourism, for example, are considered to be less vulnerable to environmental changes than NBT activities (see Brooks et al., 2020; Green & Saarinen, 2022). In general, the tourism sector in the Global North, especially in the high latitudes, is expected to benefit, and the Global South tourism sector is expected to experience adverse effects of climate change (Tervo-Kankare et al., 2018a). This division is based on the estimated direct impacts of climate change on environmental conditions in the coming decades. However, there is a wide agreement that eventually, the direct and indirect impacts of climate change, will have serious adverse effects on all human activities globally, especially the Global North and South, if climate change policies fail to reduce future climate impacts and carbon emissions. Some regions will be impacted earlier and/or with more serious changes than some other areas. In this respect, Scott et al. (2019) note that the Paris Agreement (United Nations, 2016) makes several references to particularly vulnerable countries and regions.

The question of which countries and regions are most vulnerable is high-lighted by Ford et al. (2015), for example. As noted, there is a common un-derstanding that the countries in the Global South, with limited resources, are most vulnerable (Hoogendoorn & Fitchett, 2018a; IPCC, 2007; Roger-son, 2016). Indeed, vulnerability to environmental change does not exist in isolation from people, economies, and societies (Adger, 2006), and it is also dependent on adaptive capacity, calling for climate change awareness, knowledge, and sufficient resources. While the southern African region and its tourism economy can be portrayed as vulnerable to climate change im-pacts based on scarce resources and high exposure and sensitivity, in gen-eral, there is also evidence suggesting that communities themselves have significant capacity to adapt latently in local knowledge and experience of coping with variability (Adger, 2006; Brooks et al., 2020). Therefore, while vulnerability is often a negatively loaded term, it also involves a dimension or possibility of having adaptive capacity towards changes. This connects vulnerability to resilience in the studies on tourism development and chang-ing ecological and social systems (Lew & Cheer, 2018; Saarinen & Gill, 2019).

Resilience thinking

Resilience has emerged as a key term in tourism destination and regional development studies and discussions. Sometimes, resilience is seen as be-ing the new sustainability, and the concepts of resilience and sustainable development in tourism are widely perceived as integral and complemen-tary (see Butler, 2017; Lew, 2014). Related to this, McCool (2015) has stated that it is the role of sustainable tourism to support resilience. Espiner et al. (2017, p. 6) provide a somewhat opposite view, arguing that resilience is an integral part "but not sufficient for sustainability" in tourism. Indeed, re-silience, in itself, is not necessarily a desirable state (Stockholm Resilience Centre, 2015) and does not automatically lead to sustainability in tourism; a destination system can be highly resilient, but based on a negative lock-in modes in tourism development that could, for example, work against the sustainability and well-being of nearby vulnerable communities (Simmie & Martin, 2010, p. 32; Saarinen & Gill, 2019). Studies in southern Africa have demonstrated that many successful adaption mechanisms to address cli-mate change impacts, such as the increasing use of air-conditioners in tour-ism facilities, can contribute to the resilience of the businesses, but turn out to be problematic from the perspective of sustainability management in tourism (Hambira et al., 2013; Saarinen et al., 2012).

In general, resilience refers to how a system or unit, such as a tourism destination, regional business network, or single business, responds without collapsing due to external changes, pressures, and shocks (see Hall et al., 2018; Saarinen & Gill, 2019). In his seminal work, Holling (1973, p. 17) de-fined resilience as "the ability of these [natural] systems to absorb changes of state variable, driving variable, and parameters, and still persist". While the concept originally focused on the stability of ecological systems and how ecosystems reacted to change, disturbance, stress, and other events, resilience thinking relatively soon began to influence research beyond ecological studies (see Davidson, 2010; Davoudi, 2012; Folke, 2006). Today, the idea is widely used in wider system analyses including community resilience and tourism resilience (see Butler, 2017; Calgaro et al., 2014; Espiner & Becken, 2014). Community resilience, in particular, has been studied in southern Africa, but rarely in tourism development contexts (see Saarinen et al., 2020).

Folke (2006) has divided resilience approaches, based on their scope, into three different types of resilience. The first, *engineering resilience*, represents a focused and technical interpretation of resilience-building. It emphasises the efficiency of a system's ability to return to a steady state (equilibrium) that was perceived to exist before an external change, disturbance, or shock took place. Typically, this is used in relation to natural hazards and disasters (Hall et al., 2016) such as floods and heat-waves. In the current context of climate change thinking, these incidents are understood to be caused or significantly worsened by human actions, making those events anything but natural. The engineering term relates to approaches and techniques that strengthen infrastructures (e.g. buildings, dams, and tourism-related facilities such as harbours), which are increasingly used in the coastal areas of southern Africa (see Brooks et al., 2020; Fitchett et al., 2016c; Hoogendoorn & Fitchett, 2018a).

In Mauritius, for example, beach erosion has become a serious problem for coastal tourism due to accelerating SLR (Anisimov et al., 2020) (Box 6.2).

Figure 6.1 Beach erosion at the Flic-en-Flac seaside tourist resort, Mauritius. Photograph by Jarkko Saarinen.

This has been noted to have a negative impact on the attractiveness of coastal tourism destinations and beach-related activities (Perch-Nielsen, 2010). The beach erosion has been managed by physical systems such as sandbags (Figure 6.1). However, based on the study by Duvat et al. (2020) on 60 beach sites in Mauritius over the period 1999–2019, there is a need for a shift from a strict 'one-size-fits-all' engineering response to soft and place-specific responses that are ecologically sensitive. This relates to Folke's (2006) *ecological/ecosystem (or social) resilience*. Compared to the engineering approach, it is a broader and more dynamic way to think about and plan resilience. It focuses on the persistence and robustness of a wider ecosystem or community to relate with change and shocks. Furthermore, it acknowledges that instead of returning to a specific pre-event condition, after a shock, there can be multiple equilibria. Thus, by adapting to change, a resilient system may not return to a prior level, but different ecological or social equilibria may be typical and used for different circumstances (Lew, 2014).

The third, and probably, the most complex but realistic, approach is called *socio-ecological resilience* (Folke, 2006). It is based on the understanding of the integration of ecological and social processes, with related integral elements of adaptive capacity and transformability via learning and innovation (Saarinen & Gill, 2019). Thus, systems are not understood as either natural or social by nature nor are they considered to be characterised by stable conditions; systems constantly evolve and change even without major external stresses and shock. For example, a significant part of seasonality in natural environments and tourism is built into the systems themselves. In southern Africa, safari tourism operators effectively use environmental seasonality and related knowledge in guiding and locating wildlife for tourists to see at certain kinds of sites at different times of the year (see Manrai et al., 2020). However, as indicated by Dube and Nhamo (2020b), climate change impacts are creating critical threats to southern African biodiversity (Di Minin et al., 2016), resulting in changes in seasons and challenging the seasonal occurrence and resilience of flora and fauna. This will create major difficulties for the current and especially future safari business in southern Africa.

BOX 6.2 ISLAND TOURISM AND CLIMATE CHANGE: MAURITIUS

Jarkko Saarinen and Kiran Dookhony-Ramphul

The Sixth Assessment Report of the IPCC's (2021), reporting on the physical science basis of global climate change impacts, states that it is very likely that extreme weather events and relative SLR will accelerate in small islands and coastal regions in the future. The report indicates with high confidence that SLR results in increasing frequency and severity of coastal flooding in low-lying areas and erosion along most sandy coasts. In general, small island developing States (SIDS) are considered vulnerable to the impacts of climate change due to their limited resources and resulting low adaptive capacity (see Robinson, 2020). Indeed, the island of Mauritius, located in the South Indian Ocean tropical cyclone belt, is highly vulnerable to climate change-induced extreme events. According to the Mauritius Meteorological Services (2021), the effects of climate change have already been severe. Sea level in the southwest Indian Ocean has increased at around 1.5 mm/year at the capital city of Port Louis for the period 1950–2001. Mahadew and Appadoo (2019) have noted that climate change impacts in Mauritius over the last two decades include the lengthening of the intermediate dry season and an increasing frequency and strength of extreme weather events, heavy rains, and severe storms such as tropical cyclones.

In Mauritius, climate change-related coastal risks have a direct and significant effect on the natural ecosystems, community livelihoods, and the overall economy. The coastal areas have a very high economic value derived from coastal tourism, as over 90% of the country's hotel bed capacity is located in the coastal area. In 2019, the contribution of tourism to GDP was estimated to be well over 20% (African Development Bank Group, 2021). However, despite the importance of the (coastal) tourism sector, Mahadew and Appadoo (2019) conclude that the existing adaptation and mitigation frameworks are insufficient to address the impacts of climate change in the tourism sector. There are many individual projects that involve both public and private sectors, but with insufficient coordination between the sectors and actors along the coastline (Anisimov et al., 2020).

For example, a study by Roheemun (2018) found that almost all of the surveyed hotels (=65) had experienced negative impacts associated with extreme events caused by climate change. Specifically, over 90% of the hotels had experienced beach erosion in the past five years due to intensified swells and tidal waves. Most hotels had put in place their own structures (e.g. sandbags) to protect their beachfront from hazards (see also Anisimov et al., 2020). Based on past experiences, the hotel representatives expected that the main impacts of future climate change will be related to beach erosion, water shortages, and extreme rainfall and floods. Adaptation and mitigation plans were seen as essential components, but these are presently insufficiently developed in Mauritius. Thus, to increase their adaptive capacity, hotel representatives commonly agreed that there is still a need to develop climate change-informed knowledge, skilled employees, government finance, and governance frameworks. The absence of these critical elements hinders the practical implementation of adaptation and mitigation strategies.

In this respect, Mahadew and Appadoo (2019) highlight a need to educate and raise awareness about climate change and its effects among all those directly or indirectly involved with the tourism sector. In addition, they note a serious lack of research on the nexus between climate change and tourism in Mauritius. Considering how important the tourism sector is for the country, there is also an insufficient amount of research available to determine the best possible adaptation and mitigation strategies, policies, and plans. Furthermore, Mahadew and Appadoo emphasise that the legal and normative frameworks on climate change need to be developed in Mauritius. Similarly, Anisomov et al. (2020) call for proper governance frameworks and arrangements as the current individual ad hoc, reactive projects do not contribute to long-term coastal risk reduction and climate change adaptation and mitigation strategies, but instead jeopardise the sustainable development of tourism in Mauritius.

All types of resilience thinking have been applied in tourism (Hall et al., 2018; Lew & Cheer, 2018), but socio-ecological resilience is adopted in most tourism- and resilience-related studies (see Lew et al., 2016). This can be explained by the broader focus that goes beyond the nature-social dichotomy, as it allows thirsty scholars to consider tourism in its environmentally and socio-economically dependent context. Furthermore, a broader triple-bottom informed focus allows researchers to think about the transitional needs of establishing pathways towards sustainable development in tourism (Saarinen & Gill, 2019), involving ecological, social, and economic elements in resilience-building (see Christopherson et al., 2010; Simmie & Martin, 2010). From this perspective, change is seen as a constant state in tourism systems, and resilience indicates the capacity of a system to cope with that change and absorb disturbances by sustaining itself. This takes place by transforming with changes in tourism development as also the conditions of development are in a constant state of change (see Adger, 2000; Lew & Cheer, 2018). This stresses the importance of understanding the wider context of changing environments in tourism.

In this respect, many scholars consider the resilience of tourism systems to be closely related to the idea of destination governance (Luthje & Wyss, 2014; Saarinen & Gill, 2019). As for the idea of resilience, governance thinking also represents "the tool by which the destination adapts to change" at a system level (Baggio et al., 2010, p. 52). In general, governance refers to supporting and guiding the process of collaboration and collective action in tourism and destination development and management characterised by constant change (Bramwell & Lane, 2011; Gill & Williams, 2011). Furthermore, this change management towards resilience happens in different (organisational and spatial) scales (Luthje & Wyss, 2014). These scales are often divided into individual, organisational, and destinations resilience (Hall et al., 2018). Here, in relation to southern African tourism, we are specifically interested in the destination resilience perspective.

Southern African perspectives on destination resilience and vulnerabilities

Destination resilience "lies at the heart of much thinking about tourism and resilience" (Hall et al., 2018, p. 104). Based on this statement, it is surprising how little research, especially empirical work, has focused on climate change and pressures it creates impacting destination resilience (see Luthje & Wyss, 2014; Strickland-Munro et al., 2010). There has been increasing interest in tourism resilience studies, but this represents a quite recent focus and studies have been often limited to a certain sector of tourism activities or destination functions. The destination scale itself is problematic for empirical research. Conceptually, it can refer to a varying range of spatial scales in tourism (e.g. continents, sub-continents, states, provinces, cities/municipalities, and tourist resorts). Administrative units often provide resources and limits for resilience planning but these units such as provinces, municipalities, and their territorial boundaries are not usually meaningful for most visitors, and they are definitely not meaningful for the ecosystem attractiveness on which the destination is based (see Santarém et al., 2020, 2021). Moreover, tourism destinations are not just about tourism and tourist activities. There are also various infrastructures, properties, and private and public services that are used not only by tourists but also communities who live and work at a destination (Hall et al., 2018).

Tourism destination resilience refers to the destination system's capacity and ability to recover from changes and external shocks and continue to develop. This capacity enables the destination system "to deal with stresses by maintaining the stability of the tourism-related regional economy while ensuring the flexibility and diversity necessary for innovation and further development" (Luthje & Wyss, 2014, p. 161). Destination resilience is partly a function of individual and organisational resilience but the characteristics of the destination system itself and its internal and external networks, social capital, stakeholder relations, leadership, innovations, and resources play a major role (Cochrane, 2010; Hall et al., 2018). Therefore, the nature of tourism supply and demand is not a sufficient scope for analysing destination resilience to external changes. Instead of tourism-based elements alone, wider socio-ecological dimensions and connections at different scales need to be considered. This way, resilience thinking would have more explanatory power compared to vulnerability assessments (see Luthje & Wyss, 2014).

One of the few examples of resilience-informed studies on climate change-induced stress to tourism destinations in southern Africa is the study by Dube et al. (2020). They focused on the impacts of drought (2015–2018) on the tourism sector in the Western Cape (see Box 8.2) and the response and recovery issues for the sector and the region. The case provides a good lesson on destination resilience thinking, specifically in the South African tourism development and regional planning context. Based on their results, the climate change-triggered drought had a severe and damaging impact on the communities, businesses, and regional tourism sector resulting in the loss of jobs and revenue. With respect to vulnerability and adaptive capacity leading to destination resilience, their results highlighted that there is an urgent need for drought and change management by governments, destination developers and marketers, tourism businesses, and academia. Thus, the tourism destination needs to be understood as based on a wider network of actors than the tourism sector alone. They also noted a need for adaptive measures that could include investment in water harvest technology (e.g. air–water harvesting and seawater desalination) to ensure water independency. Furthermore, Dube et al. (2020) state that tourism must mitigate climate change to ensure the sector's resilience. This stresses long-term adaptation needs in energy and water efficiency that would lead to a drought-resilient tourism destination in the future.

While there are a limited amount studies on destination resilience, especially in relation to climate change impacts in southern Africa, this is not the case for vulnerability research. Vulnerability research is an integral part of destination resilience (see Hall et al., 2018), and the majority of vulnerability studies in tourism have also focused on water use or water-related issues in southern Africa (see Dube et al., 2020; Hambira et al., 2013; Rogerson, 2016; Saarinen et al., 2012). This is logical as the region's tourism sector and its different products and activities are highly dependent on the existence and use of water resources, whether directly or indirectly. Furthermore, many places in the southern African region are suffering from intensified water shortages and droughts due to climate change, which makes the topic well known and urgent to study. As a result, there are many demonstrative cases indicating the vulnerabilities of the sector to the impact and threats of droughts in different parts of southern Africa. Victoria Falls (both the Zambian and Zimbabwean sides), and its lessening water volume, has been widely studied (see Dube & Nhamo, 2019a). Furthermore, the impacts of the droughts on NBT in the South African national parks and game reserves have been identified in the literature (see Smith & Fitchett, 2020). According to Dube and Nhamo (2020c), droughts may also play a role in food shortages, higher energy costs, and lower visitation numbers in national parks. For example, for Lake Kariba, Zimbabwe, Dube, and Nhamo (2020b) concluded that during drought years, hotel occupancy rates were negatively impacted, and the Kariba resort had increased the managing costs due to energy shortages that also negatively affected the attractiveness of the resort.

In addition, there have been drought and water shortages related to vulnerability and adaptive capacity studies in the Kalahari region (Saarinen et al., 2012, 2020) and the Okavango Delta in Botswana (Darkoh et al., 2014; Dube, 2003; Hambira et al., 2013, 2021) and in Namibia (Tervo-Kankare et al., 2018b). The latter is the driest country in southern Africa; over 90% of the country's land area is characterised by arid conditions, with a generally hot and dry climate, characterised by sparse and increasingly unpredictable rainfall. There has been a recurrent drought since 2013, and the year 2019 was reported as the driest in 90 years of recorded history in Namibia. Due to this situation, the government has declared three states of emergencies between 2013 and 2019 (Keja-Kaereho & Tjizu, 2019). Droughts have been very problematic especially in the rural parts of the country, which are highly dependent on livelihoods based on the direct use of natural resources and the conditions of ecosystem services. Related to this, Saarinen (2016) and Green and Saarinen (2022) have concluded that due to droughts and increasing environmental change, many local communities have moved away from traditional livelihoods towards the tourism economy based on their cultural attractiveness to international tourists. This adaptive capacity has resulted in monetary income and (partial) employment opportunities, but also made communities tourism-dependent and vulnerable to new kinds of external changes that communities have no agency or power to guide or control (Kavita & Saarinen, 2016; Lapeyre, 2010). The impacts of this increasing tourism-dependency for communities and their resilience are yet unknown, but as noted by Robbins (2012, p. 23), changing environmental conditions "lead to new kinds of people" with a different set of opportunities, constraints, and threats in the future.

In addition to droughts and water shortages, scholars have studied water-based vulnerabilities and threats more broadly in southern Africa. Fitchett et al. (2016) studied the potential climate change impacts for coastal towns of St Francis Bay and Cape St Francis in the Eastern Cape Province of South Africa (see also Hoogendoorn et al., 2016). Their study focused on risks such as flooding that are induced by SLR. Giddy et al. (2017a) studied white-water sports and their vulnerability to climate variability in Namibia and South Africa, based on the experiences and perceptions of white-water tourism operators. White-water adventure tourism and sports tourism are heavily reliant on predictable river water levels, and Giddy et al. (2017a) found that extreme and unpredictable weather patterns are already impacting tourism operators' ability to provide quality services. This kind of approach, focusing on (i) tourism operators', (ii) tourists', or (iii) tourism-dependent community members' perceptions, awareness, and responses to climate change and its impacts, is widely utilised in southern African scholarship (see Hambira et al., 2013, 2021; Hoogendoorn et al., 2016; Manwa et al., 2017; Mushawemhuka et al., 2018; Pandy, 2017; Pandy & Rogerson, 2018; Saarinen et al., 2012, 2020). These studies cannot be regarded as representing tourism resilience research. However, by focusing on stakeholders' climate change perceptions, awareness, networks, and internal relations and impact and risk assessments, they involve key elements of destination resilience and an adaptive capacity-building process.

Conclusions: towards resilience governance

Resilience has become one of the key ideas and approaches in tourism and climate change research. However, there is still a shortage of destination-specific resilience studies with a wider focus on socio-ecological environments. This applies to southern Africa and beyond. Furthermore, as noted earlier, resilience itself is not necessarily desirable nor an automatically sustainable state for destination development. Thus, there is a need to expand resilience research towards sustainable development and destination governance in southern Africa. In this respect, a sustainable tourism governance approach would involve the tourism markets but also context-specific socio-ecological elements and institutional structures that go beyond tourist operators, destination management organisations, customers, and their relations. Indeed, as highlighted by Folke (2006, p. 260), there is a "fundamental importance for governing and managing a transition towards more sustainable development paths" in resilience-building (see Espiner et al., 2017). Sustainable destination governance strengthens resilience and adaptive capacity and as such could lead regional tourism development towards sustainability in the era of climate change. The key challenge is implementation, which requires efficient and SDG-informed climate change policies guiding future southern African tourism development.

7 Adaptation to climate change by the southern African tourism sector

Introduction

Defining adaptation to climate change is challenging and often fails to include and describe the complete gambit differences, contingencies, and local contexts across the globe that shape the adaptation strategies and their capacity for success. Perhaps one of the best definitions, which attempt to address these complexities, is presented by Smit and Pilifosova (2001, p. 879), who define adaptation to climate change as:

> …adjustments in ecological, social, or economic systems in response to actual or expected climate stimuli and their effects or impacts. It refers to changes in processes, practices, and structures to moderate potential damages or to benefit from opportunities associated with climate change.

Indeed, the 2008 report of the United Nations World Tourism Organization (UNTWO) made it very clear that the impacts of climate change will be a severe threat to the potential sustainability of tourism sector throughout the 21st century (UNWTO, 2008). A range of adaptation strategies have been implemented within the tourism sectors of the Global North, particularly the construction of more resilient transportation infrastructure, on which the tourism sector depends. Examples include increased artificial snowmaking capabilities for ski resorts (Steiger & Scott, 2020), SLR strategies such as the building of defence barriers, and the physical movement and replacement of infrastructure (Schliephack & Dickinson, 2017). However, Adger et al. (2003) note that the level of adaptive capacity differs across the world, with the most limited capacity in the Global South. This is certainly the case in southern Africa, where few tangible adaptation strategies have been implemented, and little focus has been placed on adaptation in the tourism sector (Hoogendoorn et al., 2021).

One way of approaching adaptation and the development of robust adaptive capacity is through adaptive co-management (Armitage et al., 2008). An adaptive co-management approach is where "…flexible community-based systems of resource management tailored to specific places and situations

DOI: 10.4324/9781003102618-7

and supported by, and working with, various organizations at different levels" (Olsson et al., 2004, p. 75). Adaptive co-management is often implicitly enacted within the tourism sector, facilitating different levels of stakeholder engagement, allowing particularly marginalised groups to be involved in the governance of resources and the development of local and attraction-specific adaptation plans (Islam et al., 2018; Lai et al., 2016; Plummer & Fennell, 2009). However, this has yet to be formally applied as a deliberate approach in driving the development of adaptation plans within the tourism sector nor has it been explicitly reflected on as a theoretical position in the literature on tourism or the tourism climate nexus. This is despite repeated calls for improved understanding of community views on climate change and its relation to tourism, which would allow for "contextual adaptation challenges to be met in a more sustainable way" (Kaján & Saarinen, 2013, p. 167), and the growing body of literature on adaptive co-management in the climate change research domain (see Plummer, 2013; Westskog et al., 2017; Wilson et al., 2018).

Throughout this book, the lack of adaptive capacity in the Global South is often mentioned. For this chapter, it is important to define what is meant by the term 'adaptive capacity'. These definitions also vary by context, but for this book, we consider the definition of Smit and Wandel (2006, p. 282): "In numerous social science fields, adaptations are considered as a response to risks associated with the interaction of environmental hazards and human vulnerability or adaptive capacity". This implies that the response is dependent on the level of human vulnerability and the capacity to adapt to risks that emerge. Adaptive co-management, in this context, places the local risk context and human vulnerability at the forefront and allows for a community-based solution-driven approach to climate and environmental hazards, in a manner which is feasible in the context of simultaneously addressing a range of other urgent infrastructural and livelihood needs (Fitchett, 2014). In a sector dominated by SMMEs, much of the adaptation that has been implemented thus far could be considered to have been selected through a process of, at least partial, adaptive co-management, and this framework would provide a useful pivot for future efforts particularly as the climate change threats start to require municipal-scale rather than property-scale adaptation (Baird et al., 2016).

To identify current adaptation strategies utilised in the southern African tourism sector and to reflect on future strategies, this chapter is divided into three sections focusing on large-cost infrastructural adaptation, medium, and low-cost infrastructural adaptation and non-infrastructural adaptation to climate change in the southern African context.

Larger-cost infrastructural adaptation

Large-scale, often also large-cost, adaptation mechanisms are arguably the most effective approach in reducing the risk of future climate change (Figure 7.1). There are several examples from the Global North where

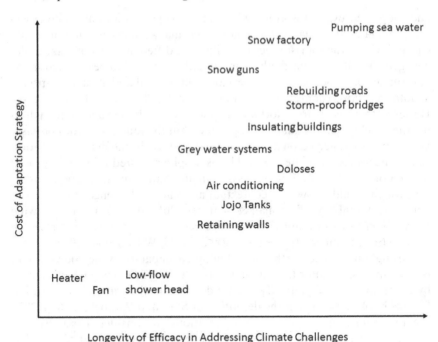

Figure 7.1 Costs of adaptation strategy versus the longevity of efficacy.

countries are investing in extremely expensive major large-scale infrastructural developments. By the early 2000s, major approaches had started, including the removal and rebuilding of infrastructure such as a roads, raising the height of bridges, making use of intensive pumping to avoid saltwater intrusion, large-scale sustainable land management practices, and the development of storm surge barriers and dyke systems (Eekhout & de Vente, 2019; Kabat et al., 2005; OECD, 2018; Rosenzweig et al., 2011). At the opposite extreme, in the case of the Boyne catchment in Ireland (Murphy et al., 2011) and glaciers in South East Iceland (Rutty et al., 2021), a 'wait-and-see' strategy is being implemented, holding off on any infrastructural adaptation until the trajectory of climate change and the veracity is confirmed by empirical data. For southern Africa, many of the risks of climate change have been highlighted in recent years; yet, large-scale infrastructural adaptation remains difficult to implement given the large capital requirements. However, there are a few clear examples where individual municipalities have made tangible progress.

Adaptation measures in the City of Cape Town

The City of Cape Town municipality started thinking about adaptation and mitigation mechanisms as early as 2001 with formal government appointments made during 2005 (Mukheibir & Ziervogel, 2007). Indeed, Ziervogel and Parnell (2014) note that one of the successes of the City of Cape Town's adaptation strategy is that different stakeholders from different spheres, including NGOs, scientists, government, and the private sector, have congregated to form the so-called *City of Cape Town Climate Change Think Tank*. This formed a collective sense of responsibility in planning for the future impacts of climate change. However, Ziervogel and Parnell (2014) also note that, unfortunately, climate change adaptation in the City of Cape Town functions most effectively when there are specific crises facing the municipality rather than through longer-term planning. The slow pace of cooperative adaptation is considered by Ziervogel et al. (2010), who argue that to avoid disastrous consequences for a city that has high levels of informality, a large tourism sector and a water-hungry agricultural sector surrounding the city, adaptation should be significantly more urgent.

The City of Cape Town municipality is situated in the drier western half of the country (Jordhus-Lier et al., 2019). The metropolitan region has already had a taste of the impacts of climate change. Severe water scarcity was experienced during the 'Day Zero' drought, culminating in early 2018 (Enqvist & Ziervogel, 2019). This drought ramped up adaptation measures from both the private sector and public sector, albeit not necessarily in collaboration with each other (Matikinca et al., 2020). Approaches included the controversial revoking of universal free basic water provision and fixed charges implemented, plans for the development of desalinisation plants, household flow regulators, and upscaling reuse schemes. Quality service provision, including fixing leaks in water pipes across the city, was successful in addressing water shortages, as were marketing campaigns to create awareness of water scarcity.

Private citizens who could afford to drill boreholes did so during the drought, albeit contributing to a further reduction in the groundwater level, while more eco-friendly homeowners installed rainwater tanks and greywater systems instead (Robins, 2019). Removal of water-thirsty invasive plant species and the redevelopment of water-wise gardens were undertaken on both private and public land. At the height of the drought, even more extreme geoengineering strategies were touted, such as towing icebergs from Antarctica to provide fresh water (Malan, 2018). The commitment from different economic sectors to use water responsibly and sparingly was of great importance and identifying and managing water-stressed zones (Millington & Scheba, 2020; Ziervogel, 2019). Prinsloo (2019) found that in the case of the sharing economy in Cape Town, Airbnb hosts made a concerted effort to address their water consumption during the drought but also put in place water saving and water reduction mechanisms for future droughts. It seems that Airbnb hosts are generally aware that droughts in future will increase in severity and regularity in future. In terms of the Cape Town tourism sector at large, Dube et al. (2020) argue for strong partnerships, which echo the approach of adaptive co-management, to assist the tourism sector at large to adapt to future threats of drought, which Pascale et al. (2020) argue are likely to increase in frequency and intensity.

Adaptation measures in the eThekwini municipality

One example of a South African municipality that is planning for future climate change threats, broadly and specific to infrastructure on which the tourism sector depends, is the eThekwini Municipality which includes the major city of Durban (Roberts, 2010; Roberts et al., 2020). The 2020 Integrated Development Plan (IDP) is an example of forward thinking regarding climate change adaptation at municipal level. The IDP was developed in consultation with different stakeholders through workshops, discussions, and gathering specialist and indigenous knowledge. The eThekwini IDP provides a far more concrete approach to adaptation planning than can be seen in many of the other major metropolitan centres in the country and was developed in line with the projections of the IPCC assessment reports and the commitments to the Paris Agreement. This is especially important for the city of Durban which is well known as one of the major sun, sea, and sand and urban tourism destinations in the country, especially for domestic tourists (Preston-Whyte, 2001; Preston-Whyte & Scott, 2007).

The eThekwini IDP highlights the possibility of economic decline resulting from climate change, causing a decline in infrastructure and infrastructural upkeep, especially for coastal properties, which could lead to general loss in revenue (eThekwini Municipality, 2020, p. 203). The eThekwini Municipality established the Municipal Climate Protection Programme in 2004 to formally consider responses to climate change concerns, and in 2015 different divisions of the eThekwini Municipality came together to form the Durban Climate Change Strategy (DCCS) to address both mitigation and adaptation. This strategy focuses on five specific adaptation themes, namely, biodiversity, water, health, food security, and SLR (eThekwini Municipality, 2020, p. 88). Tangible examples of the outputs from this strategy are the development of early warning systems, innovative informal settlement designs, and planting trees to stabilise soil, especially in areas with steep slopes that are common in the greater metropolitan area (eThekwini Municipality, 2020, p. 175). The IDP highlights the need to build a city that is resilient to climate change. The ideal of this strategy is a well-run municipality that is well adapted to the threats posed by climate change and will allow tourism business to focus their efforts on long-term investments rather than more immediate and potentially localised responsive adaptation measures. It will be valuable to reflect on the effectiveness of the IDP and its implementation in years to come, and the degree to which it can be viewed as effective adaptive co-management both in the tourism sector and more broadly.

Adaptation in smaller urban areas

In terms of smaller urban areas, such as coastal towns, the example of adaptation to climate change within the tourism sectors of St Francis Bay and

Cape St. Francis has been documented (Fitchett et al., 2016a; Hoogendoorn et al., 2016). These two towns make extensive use of the 'dolos', a South African-invented system of interlocking concrete blocks, to protect against storm surges and the impacts of SLR (Hoogendoorn et al., 2016). This local solution for a local problem emphasises the need for driving innovations and responses to climate change threats, rather than waiting for technology to be developed in the Global North, and struggling to finance these. However, the question remains as to how tourism operators across the southern African region can leverage clear and effective adaptation measures to address all aspects of the changing climate and how the design of tourism infrastructure should be developed for future climate. Mather and Stretch (2012, p. 251) are of the view that in many cases, especially when it comes to urban tourism amenities that have been developed over many decades, it is not practical to 'retreat and reconfigure' these amenities, and thus that only a small range of adaptation measures are ultimately possible.

BOX 7.1 ADAPTATION OF TOURISM INFRASTRUCTURE FOR CLIMATE CHANGE

Anne Fitchett

Tourism infrastructure ideally needs to be perfectly comfortable every day: owners of tourist venues need to provide amenity even when the weather prevents outdoor activities. When the weather is not suitable for the activities that have brought visitors to the locality, the accommodation has to provide alternative activities in a comfortable setting. People's perceptions of comfort are varied (Buso et al., 2017), so it is advisable to allow each room to be controlled separately for heating, cooling, and ventilation.

In warm temperate regions, orientation is the most economical way of ensuring comfort (Morrissey et al., 2011). In the Southern Hemisphere, north-facing rooms can have well-proportioned windows and simple shading devices that block solar radiation in summer, yet allow ingress in winter (SANS204, 2011). East and west-facing openings should be as narrow as possible or have wide pergolas or verandas. Except in very warm localities, south-facing openings should be small and well-insulated. In many tourist venues, ideal orientation may not offer the best view, especially where the tourist attraction is intimately associated with the view, such as beach or wildlife tourism (Alonso & Ogle, 2008). This can be overcome by designing a new amenity, such as a gazebo.

Some parts of southern Africa experience extreme temperature, and even those in the more temperate regions have short periods of excessive heat or cold (van der Walt & Fitchett, 2021a). The most cost-effective way of addressing temperature variation is through insulation, especially in the roof-space, to prevent heat buildup on the most exposed surface and to prevent warmth generated within a space from escaping through the ceiling (Jelle, 2011). A flat roof can be insulated with ceiling panels or can have a roof-garden that moderates the temperature beneath it (Fitchett et al., 2020a).

In hot climates, a number of devices have evolved throughout the world (Zhou et al., 2008). Excellent ventilation is achieved by having openings in at least two walls of a room. Breezeways, in the form of a passage open at both ends, induce suction and benefit all rooms opening onto it. Verandas have a similar function as well as providing cool sheltered spaces.

In very cold climates, all openings should be as small as possible, tightly sealed, and protected with heavy curtains, padded blinds, or shutters. New buildings in extreme climates should ideally be very compact to prevent excessive heat gain or loss, but may compromise aspect, so insulation and ventilation are cost-effective alternatives (Su, 2011).

Tourist facilities are often located close to the tourist attractions, such as a natural water body. This increases the vulnerability of the buildings, especially to extreme climate events such as tide surges and flooding. New buildings can be elevated above the natural ground line, raised on stilts or above service rooms (Mycoo, 2014). Existing buildings can be protected by appealing features such as vegetated berms or low hedges.

Sustainable Drainage Systems (SuDS) attenuate storm-water as well as providing amenity such as a detention pond that doubles up as a sports court in dry weather (Fitchett, 2017). Green roofs also moderate storm volume and intensity. The extreme wind and rainfall events that can be expected with climate change can be devastating on tourism infrastructure, especially when located near water features. Retaining walls can provide a robust separation of buildings from the natural feature and can be combined with grading and terracing. A combination of paving, graded areas, vegetation, and SuDS can optimise the resilience to extreme climate events, while providing amenity for visitors (Woods-Ballard et al., 2007).

Lastly, buildings, infrastructure, and landscape all need to be reviewed regularly to ensure that they are able to resist extreme events. In a severe storm, un-secured elements are ripped away, roofs are lifted, and projecting elements are overturned. Buildings that have deep eaves or verandas are particularly susceptible, so they need special care to ensure that they are tied down. In highly vulnerable areas, there may need to be a collaboration of individual property owners and official bodies to motivate for the addition of large-scale interventions such as dolosse to serve as a sea-wall (Schneider, 2014).

An example of less than effective proactive adaptation at municipal level was seen in the case of the Mopani District in South Africa following floods caused by Tropical Storm Dando, as mentioned in Chapter 5. In the absence of effective forward planning for climate change and municipal-level infra-structural improvement, the road and bridge networks suffered significant damage, with long-term impacts on access to destinations and accommo-dation (Fitchett et al., 2016a). Because of the limited capacity to deal with such an extreme event by government, private citizens developed their own adaptation mechanisms such as constructing temporary roads and commu-nicating with rescue personnel for the evacuation of individuals isolated by the floods (Fitchett et al., 2016c). Of the 24 tourist accommodation estab-lishments that were interviewed in 2012, the direct costs of the storm were estimated at R58.92 million, and costs relating to loss of business were in the region of R4.230 million (R8.69 = US$1 in 2012; Fitchett et al., 2016c). It was also foreseen by respondents that costs of future adaptation would exceed hundreds of thousands of Rands. These cost estimates were limited to the respondents of the study, while the broader impact on the tourism sector was significantly higher, over, and above costs that were never determined. Indeed, the long-term impacts of infrastructural recovery such as roads, removal of obstructions, and reestablishment of water sources continued to damage the tourism sector of the region for several years afterwards (Fitch-ett et al., 2016c).

Given the rather mixed results and planning attempts by two of South Africa's major cities, small coastal towns such as St Francis Bay and Cape St Francis, and regional municipalities such as Mopani District Municipality, much more focus should be placed in encouraging municipalities to build and design for a future under climate change. A focus on building resil-ience to climate extremes is key, especially for South Africa's other major cities such as Johannesburg, Pretoria, East London, and Gqeberha, which would require major and well-implemented strategies for tourism. Focus should also be placed on South Africa's middle order or secondary cities and smaller towns that may not have the finances and adaptive capacity that the major metropolitan cities have.

Medium and low-cost infrastructural adaptation

For individual tourism operations, medium and low-cost infrastructural adaptations are the most viable option in privately implementing adapta-tion strategies, whether in addition to or in the absence of a larger-scale municipal response (see Hallegette, 2009; Hoogendoorn & Fitchett, 2018b). These approaches seldom address the low-frequency high-intensity ex-treme climate events, but rather seek to most effectively address the slower and more progressive reduction in climatic suitability of their destination as temperatures increase, rainfall patterns change, and wind and humid-ity change through low- to medium-cost equipment within their private properties. The most frequently implemented adaptation strategies include air-conditioning, heaters and fans, rainwater harvesting, and water-saving mechanisms. These examples will be discussed in this section of the chapter.

Air-conditioning

In southern Africa, air-conditioning is not common in accommodation establishments or tourism attractions. This is due to the cost of the air-conditioning units, the electrical costs incurred in operating air-conditioners, and challenges of regular electrical outages due to pressure on the electrical grid (Mushawemhuka et al., 2020). For example, Mushawemhuka (2021) found that the high-end tourism attractions in the warmer regions of Zimbabwe most commonly adapt to sweltering temperatures by using air-conditioning. Likewise, tour guides hire air-conditioned safari trucks in times of extreme heat, while bungee jumping operators make use of air-conditioned helicopters and air-conditioned briefing rooms for tourists before they take part in activities, and the more expensive hire-boats are air-conditioned to ensure the comfort tourists (Mushawemhuka, 2021). At a lower cost-point, both in terms of installation and longer-term running costs, Mushawemhuka (2021) reported that establishments are increasingly making use of thatch as it is considered cooler than other types of roofing materials. The same establishments were also found to have planted certain species of shrubs and trees to increase shade and keep areas cooler. Where new infrastructure was being built, this was preferentially sited on south-facing slopes that would receive less direct sunlight.

BOX 7.2 TOURISM ADAPTATION TO CLIMATE CHANGE IN ZIMBABWE

William Mushawemhuka

Zimbabwe is a landlocked country in southern Africa, well known for its diverse wildlife, dramatic landscape, and favourable weather for NBT. The tourism sector is the second largest contributor to the Gross Domestic Product (GDP) of the country, owing to the popular outdoor and NBT destinations, including the Victoria Falls, Hwange National Park, Gonarezhou National Park, Kariba Dam, the Great Zimbabwe Ruins World Heritage site, and the Eastern Highlands Mountain Ranges (Zibanai, 2018). Weather plays a crucial role in tourists' comfort and their overall experiences at the various tourist attractions across the country (Dube & Nhamo, 2020b; Mushawemhuka et al., 2018, 2020). The Zimbabwean tourism sector faces threats from climate changes, with projections of temperature increases of about 1.5°C by 2050 and precipitation decreases (IPCC, 2018). These shifts in climate are projected to result in the deterioration of the natural environment, the depreciation of NBT resources, and the reduction in the climate suitability of the country for tourism (Dube & Nhamo, 2020b; Mushawemhuka, 2021). Due to these climatic shifts, there are various adaptation strategies that are being employed mostly by the tourism stakeholders around the country.

With mean annual TCI scores ranging from 75.5 to 86 for the period 1989–2014, Zimbabwe is classified as having 'very good' to 'excellent' climatic suitability for tourism (Mushawemhuka et al., 2020). Despite these positive TCI results, Zimbabwean NBT destinations are already facing climate-related challenges which require urgent adaptation strategies (Mushawemhuka, 2021). These adaptation strategies can be separated into strategies for tourists and those for tourism service providers. For the tourists, the most important adaptation strategy involves coping with the various meteorological effects faced at destinations. In Zimbabwe, at destinations such as Victoria Falls, Kariba, Hwange National Park, and Gonarezhou, which are located in the low-veld; tourists experience high temperatures, yet are willing to endure the thermal stress because they perceive that the outdoor NBT activities would be of greater value to them. Tourists also have the autonomy to suspend or postpone activities or trips when adverse weather conditions occur. In Manicaland, tourists postponed their activities, waiting for suitable weather conditions. In the low-veld, tourists avoid peak temperatures at midday and carry out activities at night and or in the morning.

For the tourism stakeholders, such as accommodation establishment managers and tour operators, adaptation strategies to current and projected climate change threats vary based on the location and tourism activities. The adaptation strategies are predominantly adopted at a local level by the stakeholders at their respective locations (Mushawemhuka, 2021). For instance, most water activities along the Zambezi River, such as white-water rafting and boat cruises, are negatively affected by the reduction in water levels, mostly during the dry season. This reduction in water levels is then exacerbated when the Zambezi River Basin is affected by climate change-induced droughts, and these droughts are increasingly occurring regularly in southern Africa (Baudoin et al., 2017). Therefore, the adaptation strategies at destinations within the Zambezi River Basin (Victoria Falls and Kariba) are designed around the water-level fluctuations along the Zambezi River and Kariba Dam. These strategies include avoiding some areas along the Zambezi River during white-water rafting or completely suspending the activity when water levels are too low. In

Kariba, most shallow parts are avoided in the Kariba Dam due to fear of colliding on rocks.

For Hwange Gonarezhou National Parks, adaptation strategies are centred around the main tourist activity in this area, safari game viewing. Due to the extremely high temperatures experienced during summer, operators now offer air-conditioned safari trucks at an additional fee. Cheaper, open air safaris trucks are used for game viewing in the early mornings and evenings when temperatures are cooler (Mushawemhuka, 2021).

The tourism sector in Manicaland Province faces climate-related challenges in the form of floods and tropical cyclones, which are becoming more intense (Fitchett, 2018). In March 2019, the Province was affected by Cyclone Idai, which resulted in a huge death toll, extensive infrastructural damage, and loss of agricultural produce. The Zimbabwean government and tourism stakeholders are struggling to repair these damages, and stakeholders in the area fear that recovery will take a very long time, while other operators have had to close operations permanently (Mushawemhuka, 2021).

Notably, the Zimbabwean NBT sector is struggling to withstand the current climate change impacts and to prepare for climate change (Mushawemhuka, 2021; Mushawemhuka et al., 2018). This is evident in the country where, in most cases, the stakeholders can only manage to adapt to some selected impacts. The inability to completely circumvent all the climate change impacts results from the limited access to technology, information and, most importantly, capital to finance adaptation strategies (Saarinen et al., 2020).

For accommodation establishments in South Africa, air-conditioning is likewise not ubiquitous. To illustrate this, a database was constructed from the *Tourism Grading Council of South Africa* (TGCSA) website (www.tourismgrading.co.za) for 18 of South Africa's tourism destinations for which previous studies on tourism and climate change have been conducted (see Fitchett et al., 2017). Arguably, the tourism accommodation establishments listed by the TGCSA are among the most well-developed and established in the country, as they need to comply with a minimum set of standards to receive a grading. However, across each of the cities explored, a large proportion of accommodation establishments across each of the star gradings indicated in their booking sites that they did not have air-conditioning (Table 7.1).

The availability of air-conditioning does depend significantly on the location (Table 7.1). In warmer regions of the country, including Kimberley,

Table 7.1 Name, rating, and air-conditioning availability in accommodation establishments

	Location	1 Star	2 Star	3 Star	4 Star	5 Star	Total
Sample size	Johannesburg	1	9	25	29	9	73
Full AC		1		8	22	8	39
Partial AC							
No AC			9	12	4	1	26
Not disclosed				5	3		8
Sample size	Cape Town	1	1	52	213	61	328
Full AC		1		24	170	61	256
Partial AC				1	6		7
No AC			1	26	34		61
Not disclosed				1	3		4
Sample size	Durban	2	2	24	26	4	58
Full AC			2	21	20	4	47
Partial AC					3		3
No AC		2		3	3		8
Not disclosed							
Sample size	Polokwane		2	27	19	3	51
Full AC				18	14	2	34
Partial AC				1	1		2
No AC			2	8	4		14
Not disclosed						1	1
Sample size	Pilanesberg		1		1	1	3
Full AC					1	1	2
Partial AC							
No AC			1				1
Not disclosed							
Sample size	Kimberley	1		18	15	2	36
Full AC		1		15	12	2	30
Partial AC					2		2
No AC				3	1		4
Not disclosed							
Sample size	Port Nolloth			7			7
Full AC				1			1
Partial AC							
No AC				6			6
Not disclosed							
Sample size	Paarl			5	16	3	24
Full AC				4	15	2	21
Partial AC					1		1
No AC				1		1	2
Not disclosed							
Sample size	Knysna			21	45	21	86
Full AC				4	20	16	40
Partial AC				1	4		5
No AC				16	21	5	42
Not disclosed							
Sample size	Gqeberha	2	3	47	61	4	117
Full AC		2		25	31	4	62
Partial AC			1				1
No AC			1	21	29		51
Not disclosed				1	1		3

(Continued)

	Location	1 Star	2 Star	3 Star	4 Star	5 Star	Total
Sample size	East London	1		57	61		119
Full AC		1		20	38		59
Partial AC				1	4		5
No AC				34	17		51
Not disclosed				2	2		4
Sample size	Bloemfontein	1	2	20	26	3	52
Full AC		1		14	25	3	43
Partial AC			2	4			6
No AC				2	1		3
Not disclosed							
Sample size	Bethlehem/			11	11	1	23
Full AC	Clarens			3	6		9
Partial AC					1		1
No AC				8	4	1	13
Not disclosed							
Sample size	Ladysmith			3	5	3	11
Full AC				3	5	3	11
Partial AC							
No AC							
Not disclosed							
Sample size	St. Lucia			10	26		36
Full AC				9	26		35
Partial AC							
No AC				1			1
Not disclosed							
Sample size	Mbombela	1	2	10	24	7	44
Full AC		1	2	8	22	5	38
Partial AC							
No AC				1	2	1	4
Not disclosed				1		1	2
Sample size	Secunda/	2		5	7		14
Full AC	Ermelo	1		1	1		3
Partial AC							
No AC		1		4	6		11
Not disclosed							
Sample size	Pretoria		9	84	124	22	239
Full AC			1	44	92	21	158
Partial AC				2	5		7
No AC			4	32	20	1	57
Not disclosed			4	6	7		17

Mbombela, Ladysmith, St. Lucia, Paarl, and Durban, air-conditioning is more commonplace. Locations in cooler parts of the country, including Bethlehem and Clarens, Secunda and Ermelo, Knysna, East London, and Gqeberha, had fewer accommodation establishments with advertised air-conditioning. Well-established tourism destinations in the key economic hotspots of the country, including Johannesburg, Pretoria, and Cape Town, had a higher prevalence of air-conditioning regardless of the climatic conditions of those cities.

Invariably, the higher the star grading of the accommodation establishment, the greater the likelihood of the availability of air-conditioning. The majority of establishments in the 4–5-star range had air-conditioning in all or some of the rooms, regardless the location or climate. It is, however, notable that the 3-star accommodation establishments had the lowest prevalence of air-conditioning available in accommodation establishments. As TGCSA establishments are generally better equipped than non-graded establishments, it can be argued with confidence that the majority of 1-, 2-, and 3-star establishments in broader southern African region would not have air-conditioning. It is also important to note that air-conditioners are mostly used for cooling in southern Africa rather than as 'heat-pumps' as they are in the Global North. Instead, in the cool regions of southern Africa oil or gas heaters, electric blankets or fireplaces are used, which address the chill but do not provide constant temperature. Most accommodation establishments in southern Africa do not have double or triple glazing or insulation as effective as in much of the Northern Hemisphere, and therefore indoor temperatures are rarely consistently regulated by the building itself.

Rainwater harvesting and water saving mechanisms

Efforts to capture and store rainwater, recycle greywater, and minimise water use sit at the intersection of climate and environmental change adaptation and mitigation and are often viewed as part of efforts to green accommodation establishments (Hoogendoorn et al., 2015; Prinsloo, 2019). In the case of greening transitions undertaken at guest houses in two of South Africa's major provinces, namely the Gauteng and KwaZulu-Natal Provinces, the primary goal in implementing changes was to reduce water and electricity costs and for accreditation with the green grading association (Hoogendoorn et al., 2015). However, each of these infrastructural changes served to reduce the dependence of these tourism accommodation establishments on regional water supply and would allow them to continue to provide guests with the opportunity to shower and use other water-based amenities during water outages or during water restrictions in drought periods. These adaptations became far more commonplace in Cape Town during the 'Day Zero' drought, as mentioned previously.

While small-scale adaption strategies are being implemented at the scale of individual accommodation establishments, more effective longer-term solutions are often lagging. For example, the major metropolitan centre of Johannesburg has limited rain-harvesting systems in place, and rather, water is sourced from the Lesotho Highland Water Project, an extensive system in the adjacent country of Lesotho, more than 350 km away (Mwangi, 2007). In the wealthy northern suburbs of Johannesburg where there are numerous high-end hotels, boutique hotels, guesthouses, backpackers, and other accommodation establishments, few have installed gutter systems that channel rainwater into tanks (commonly known as 'JoJo' tanks in South Africa).

It could be argued that the tourism sectors in the wealthy cities of at least South Africa have the resources to develop such infrastructural adaptations (Rogerson & Sims, 2012). This is already commonplace in many arid areas across the globe such as Australia (Alim et al., 2020). The lack of adaptation in this regard therefore speaks to a level of apathy or misconceptions regarding the severity of the potential for water crisis during drought conditions.

The Vaal Dam, which serves as the primary reservoir for the Gauteng region, reaches critical levels every year after the dry winter season and is proverbially 'bailed out' by summer rains (Archer, 2019). Rather than risking the potential of not being able to provide tourists with water during their stay at accommodation establishments, a simple adaptation mechanism of rainwater harvesting can be applied. While there is an upfront cost in the purchase and installation of these JoJo tanks, the long-term water savings and capacity for water security serve to pay this off. At a much lower-cost, small-scale adaptation such as low-flow shower heads reduce the strain that the tourist sector places on water supply and security, in both large and heavily populated urban centres and in rural communities with water supply challenges. This demonstrates the capacity for these lower-cost local-scale adaptations to be implemented; yet, they often remain a response to a local disaster rather than a proactive approach to minimise the effects of future climate change. This is, in part, due to the non-negligible costs, which are difficult to justify when growing an SMME with more immediate financial obligations (Booyens & Visser, 2010).

Non-infrastructural adaption to climate change

Non-infrastructural adaptation could be defined as diversifying offerings and strategically advertising for optimal climatic seasons rather than building infrastructure to alter the local climate (see Gössling et al., 2012). Arguably, a non-infrastructural adaptation may be a more viable approach in many cases in southern Africa, especially where financial resources are limited.

For example, in the case of the Afriski Mountain Resort in Lesotho, adaptation to a warming climate and the resultant shortening of the ski season has included a diversification in the range of activities on offer. An emphasis is placed on non-snow-related activities during the summer, so that off-season tourists can *make the most of summer* with activities such as fishing, paintball, music festivals, enduro biking, mountain biking, hiking, abseiling, conferencing, and teambuilding, and the resort can benefit from an off-season income stream (Hoogendoorn, 2014; Hoogendoorn et al., 2021). These are all activities that are not as dependent on the narrow range of climatic conditions necessary for snowboarding and skiing. During the snow season, occupancy and visitation rates are maximised through booking long in advance and to full capacity to make optimum profit through the short ski season. A similar approach is being used for game farms in Zimbabwe. When temperatures become too hot for game drives, tourists have the opportunity to engage in other types of NBT activities such as indoor visits to crocodile farms and snake parks (Mushawemhuka, 2021). In cases where tourists are constrained to a single location, on-site diversification of offerings is important.

BOX 7.3 FUTURE OF SNOW-BASED TOURISM IN LESOTHO

Lara Stockigt

The landlocked Kingdom of Lesotho is known as the 'Kingdom in the sky' with its elevation ranging from 1,400 m.asl to 3,482 m.asl (Spooner, 2014). The mountainous region in the east experiences regular snowfall in winter, facilitating skiing since the 1960s. This prompted the establishment of the mountain lodge and ski resort named Afriski Mountain Resort in 2002. This resort has a 1-km ski slope starting at 3,222 m.asl. Afriski, as one of only two ski resort in sub-Saharan Africa, is a unique tourist attraction as snow-based tourism is an unusual offering in southern Africa. The resort is facing the extreme climatic projections of changes to precipitation levels and increased temperatures, 3.4°C–4.2°C, expected in southern Africa (IPCC, 2018). These climatic changes could reduce both the occurrence of natural snowfalls and the suitability for artificial snowmaking in the region.

With a TCI score of 64 for the period 2012–2017, the Eastern Lesotho Highlands are classified as having a 'good' climatic suitability for tourism (Noome & Fitchett, 2019). Compared to scores of neighbouring South African tourism destinations in the range of 80–89, a score of 64 is significantly lower. This score, however, is not an accurate representation of Afriski as a snow-based destination. Components of the TCI such as ample sunshine hours, preferable for beach or nature tourism, can be disadvantageous to winter-based tourism activities (Noome & Fitchett, 2019).

In response to both current and projected climate change, the resort has started to implement various adaptation strategies (Hoogendoorn et al., 2021). The first is artificial snowmaking, which started in 2005. Snowmaking at this resort involves the use of automated snow guns to guarantee adequate snowfall quality and quantities and to maintain the slopes (Stockigt, 2019). Second, the management of the Afriski Mountain Resort decides annually on the start and end-dates of the skiing season based on seasonal forecasts. The introduction of alternative activities at the resort that are less dependent on snow is a third adaptive strategy that has been implemented. The diversification of activities includes hiking trails, cycling, fishing, abseiling, adventure running, festivals, a conference centre, and high-altitude training amongst others. Many of these offerings are available year-round while others are more suited to specific seasons (Hoogendoorn et al., 2021). In the summer, the ski lift is used to pull mountain bikes up the ski slope which is transformed into an endurance track. New attractions, including a 600-m zip sail, is an example of an attraction that would appeal to winter tourists while not being dependant on snow to be functional (see Afriski Mountain Resort, 2019).

While these adaptive measures are currently practiced at Afriski Mountain Resort, conditions are expected to become more extreme. The increase in surface and atmospheric temperatures and changes in precipitation could be negative for snowmaking and the accompanying snow-based activities (Stockigt et al., 2018). However, alternative attractions could potentially benefit from warmer weather and less rain, for example, music festivals. Subsequently, these adaptive practices will have to be amplified and modified to address the aftereffects of continued climatic change. In the future systems with greater infrastructure and output, such as snowfactories may become necessary to complement or replace the snow guns (Donnelly et al., 2018).

Temporal manipulation may lead to the overall shortening of the season. Alternatively, a suggestion from existing literature that could be introduced in the future at Afriski Mountain Resort is nocturnal skiing (see Campos Rodrigues et al., 2018). This solution is rooted in the notion of night-time conditions being more suited to reliable and sufficient snow cover than daytime conditions. An additional solution that could be implemented in the future at resorts like Afriski is the introduction of indoor attractions. These would incorporate temperature control and could be impervious to the possible extreme climatic conditions of the future. Tourists at Afriski themselves suggested that in the future spa facilities and other indoor attractions could be valuable additions to the resort (Stockigt et al., 2018).

As a ski lodge in the Southern Hemisphere, Afriski Mountain Resort has lower adaptive capacity than resorts in the Global North and faces extreme projections for climate changes in southern Africa. The future and longevity of snow-based tourism at Afriski Mountain Resort is a concerning reality that necessitates the aforementioned adaptive strategies (Stockigt, 2019). Snow-based tourism resorts in the Northern Hemisphere are comparatively less vulnerable and may not yet face the challenges currently experienced and projected for Afriski Mountain Resort to the same extent (Hoogendoorn et al., 2021). Nevertheless, the conditions and practices of the Afriski Mountain Resort may provide a window into the future of northern snow-based destinations and provide a blueprint of the future conditions and the adaptive strategies that northern resorts may have to implement.

To benefit from the climatic heterogeneity of the region for multi-destination tourists, southern African tourism organisations could strategically advertise different parts of the country to maximise each region and attraction's optimal climate season (Noome & Fitchett, 2021). For example, in Zimbabwe, encouraging tourists to visit to the cooler, higher-altitude destinations in summer and warmer lowland destinations in winter is an option to reduce the likelihood of tourists experiencing heat or cold stress (Mushawemhuka, 2021). In Namibia, the TCI has been used in the development of suggested seasonal tourist routes to optimise the climatic suitability for tourists in an arid setting (Noome & Fitchett, 2021). In South Africa, an example could include advertising hiking trails as an alternative attraction when the famous Table Mountain Aerial Cable Way in Cape Town (Bickford-Smith, 2009) is not operated due to adverse wind conditions.

Extreme climate events require forward thinking to plan for low-probability or low-frequency but high-intensity events. In the case of white-water rafting sector in southern Africa, a measure of realism is necessary, given the semi-arid to arid hydroclimate and the frequency of drought events already (Giddy et al., 2017a). Non-infrastructural adaptation here ranges from monitoring river water levels before and during trips and improving the training of guides and education of tourists regarding climate through to increasing marketing and increasing ticket prices (Giddy et al., 2017a). For some operators, a complete change in business model is already under consideration (Giddy et al., 2017a). Considering a major cycling event that takes place annually in the City of Cape Town which has recently struggled with strong wind conditions, Giddy (2019) suggests adaptation mechanisms that include changing the start locations, shortening the cycle route depending on the wind and weather conditions, paying attention to weather forecasts, and taking collective action with all stakeholders.

Tourism, itself, can serve as a form of adaptation for sectors at risk of climate change, in allowing for their diversification. This is particularly the case for agriculture, where farms are increasingly leveraging opportunities in tourism and hospitality to ensure a continued income stream during years where climate hazards threaten crop yields and livestock (Little & Blau, 2020). The transition from pure food production by the agricultural sector towards multi-functional product offerings, including tourism, in rural areas has been studied in South Africa under the ambit of the 'post-productivist' transition (see Hoogendoorn, 2011b; Hoogendoorn & Visser, 2011). With climate change gradually affecting business continuity across southern Africa, the post-productivist transition towards tourism can be studied in rural areas as an effective adaptation mechanism (Hoogendoorn & Fitchett, 2018b).

Conclusion

A key challenge for climate change adaptation in southern Africa and globally is in preparing for and investing in a climate future. Climate change knowledge remains limited within the tourism sector, and climate change scenarios remain abstract to much of the non-scientific community. A common perception is that climate change will not influence tourism operations within the working careers of their managers, driving a degree of apathy and a tendency to apply lower-cost and smaller adaptation strategies (Hoogendoorn et al., 2016). In the southern African region, in particular, the socio-economic development of society needs to be weighed up against exacerbating climate change, which is at best a *Catch 22* situation for most governments with low human development standards (Saayman et al., 2012). The costs of adaptation, likewise, need to be weighed up against infrastructural development to meet the basic needs of the population. However, the devastating impacts of extreme climate events highlight the need for urgent adaptation and, as has been demonstrated in the case of the 'Day Zero' drought in Cape Town, often drive more urgent adaptation (Prinsloo, 2019). Ultimately, most effective adaptation strategy would involve the collaboration of all stakeholders in the private and public sectors, for which an approach of adaptive co-management might be most effective.

8   The role of tourism in contributing to climate change

Introduction

Scientific consensus on climate change is unambiguous – anthropogenic activities are modifying the concentrations and distributions of greenhouse gases (GHGs) in the atmosphere which absorb and scatter radiant energy, which, in turn, is resulting in global scale changes in temperature, precipitation, wind, humidity, and synoptic climatologies (Oreskes, 2004). The IPCC Special Report on Global Warming of 1.5°C (Hoegh-Guldberg et al., 2018, p. 4) reflects that "human activities are estimated to have caused approximately 1.0°C of global warming above pre-industrial levels, with a *likely* range of 0.8–1.2°C", a finding reaffirmed in the Sixth Assessment Report (IPCC, 2021). The estimated contribution of tourism to global carbon emissions ranges from 4.4% to 8% (Baumber et al., 2021; Chen et al., 2018; Lenzen et al., 2018; Richie et al., 2021), with calculations varying depending on the factors included and the date of computing. Carbon emissions per tourist are calculated to have increased by 24% from 2006 to 2014 and projected to continue to increase at 3.2% per annum (Chen et al., 2018; Lenzen et al., 2018). This is not an insignificant contribution. Moreover, with the considerable current growth in emissions trends within the tourism sector, increases in the total number of tourists, and improvements in energy efficiency and emissions reductions in other major sectors, tourism could become a major proportional and absolute source of global carbon emissions in future decades (Becken, 2002; Scott et al., 2010). Thus, while much of the focus of this book has been on the threats of climate change to the tourism sector, particularly in southern Africa, it is critical to recognise tourism as both a victim and a villain of climate change (Pandy, 2017). To blame the tourism sector alone for the challenges of climate change that it is facing would be inaccurate, but as the proportional contribution of the tourism sector to global GHG emissions increases, a robust assessment of the sources of GHG emissions, and efforts to mitigate these, will become an important component in protecting tourism sectors against the rising threats of climate change-related damage.

DOI: 10.4324/9781003102618-8

The proportional and absolute contribution of tourism to GHG emissions has increased over recent decades (Richie et al., 2021). For some, this is argued to be only a temporary phenomenon, intrinsic to the early stages of development of an economic sector. The Environmental Kuznet's Curve (EKC) hypothesis is often invoked as a theoretical and econometric argument, and research empirically testing this theory forms a notable proportion of the literature on GHG emissions from the tourism sector (Akadiri et al., 2020; Tang et al., 2017). EKC theory argues that economic development will initially lead to the deterioration of the environment during the shift from pre-industrial to industrial economies, but that after a certain level of economic growth and a shift to service sector economies, further economic growth will lead to the improvement of the environment (Akadiri et al., 2019b). There is much debate as to whether the EKC applies to the growth of the tourism sector, a broadly service sector economic segment. At extremes, Shakouri et al. (2017) and Akadiri et al. (2019b) all argue that the EKC does apply to tourism and carbon emissions, with Akadiri et al. (2019b) including Mauritius in their analysis; Gamage et al. (2017) argue that it does not apply. Occupying the middle ground, Balsalobre-Lorente et al. (2020), Sun and Hsu (2019), and Tang et al. (2017) argue that EKC could apply if the increase in economic growth were to lead to a shift to low-carbon energy sources, particularly in coal-dominated economies. When applying EKC specifically to tourism, this would require the tourism sector to implement significant changes to low-carbon energy on-site or to be influential in driving policy change to cleaner power at a national level. Akadiri et al. (2019a, 2020) do recognise that EKC can only hold if tourist arrivals do not exceed the carrying capacity of the destination country; a tenuous balance for Small Island States. Perhaps, the most nuanced view is that of Tang et al. (2017), who observe a slight EKC pattern to GHG emissions in tourism sectors as they mature, but that during stages of transformation and upgrading, emissions rise rapidly and the theory does not hold.

The GHG emissions generated within the tourism sector derive predominantly from travel, energy use, waste generation, and deforestation, while excessive water use worsens the impacts of climate change-induced aridity and drought (Amusan & Olutola, 2017; Baumber et al., 2021; Scott et al., 2010). These impacts are heightened in regions reliant on coal-fired electricity generation, and on high emission vehicles, both of which characterise much of southern Africa. Research exploring the role of southern African tourism in contributing to climate change is emerging more slowly than the literature on the threats of climate change on tourism. This chapter explores the global and local discourse on the impacts of tourism on climate change, considering transport, energy use at the tourist destination, approaches to carbon footprint calculations, and methods to mitigate these emissions.

Getting to your destination: carbon footprint of leisure transport

Transport comprises the largest component of tourism-related emissions (Baumber et al., 2021). The tourism sector is implicitly responsible for the GHG emissions associated with the transport of visitors to and from their destination and between attractions within their destination (Scott et al., 2010). It has been estimated that in 2000, tourism aviation accounted for 7% of all GHGs in Europe (Gössling & Peeters, 2007). For a tourist vacation that includes air travel, it is estimated that 60%–95% of the GHG emissions from that trip will derive from the flight (Gössling & Peeters, 2007); for Barcelona, it has been calculated that 95.6% of GHG emissions related to tourism are from arrival and departure transport, particularly through aviation (Rico et al., 2019). These numbers are particularly staggering as only 2%–6.5% of the world's population participate in air travel (Becken, 2002; Gössling & Peeters, 2007). This proportion is increasing, due to an increase in disposable income, an increase in leisure time, and competition among airlines driving a reduction in airfares (Becken, 2002; Gössling & Peeters, 2007; Higham et al., 2016).

BOX 8.1 TOURIST MOBILITIES AND CLIMATE CHANGE

Bradley Rink

Mobility plays an integral role in the production and performance of tourism. Mobility underpins the subjectivity of the tourist at the same time it delivers its salient environmental consequences. The mobility of the tourist involves more than simply leaving home. Tourism mobilities are multiple and complex, entangled across multiple scales, materialities, and meanings (Hannam & Butler, 2012). Tourism mobilities refer to myriad constellations of movement and circulation, as Sheller and Urry (2004, p. 1) note that:

> "Mobilities of people and objects, airplanes and suitcases, plants and animals, images and brands, data systems and satellites, all go into 'doing' tourism."

These multiple assemblages of tourist mobilities encompass travel to the destination, travel within the destination, as well as special 'mobile' activities such as bus tours and scenic helicopter flights (Rink, 2017) around the destination. All these movements and circulations are intrinsic to the experience of tourism, while also contributing one of its principal environmental impacts: the production of CO_2 emissions leading to climate change.

One of the primary concerns with respect to tourism mobilities and climate change is flying (Duval, 2013). Climate science has shed critical light on the impact from long-haul flying on the earth's ability to maintain radiative balance in the troposphere. It is at that altitude where long-haul aircraft establish the 'cruise' phase of flight and where they directly deposit their emissions (Unger, 2011). The combination of atmospheric science and tourism geographies is thus contributing to a more informed picture of the impact of aviation on climate change. For a country such as South Africa, located at the southern tip of Africa, the geographical distance from its predominant source markets compounds this problem. Considering the top three source countries for South African inbound international tourism in 2019 were the United Kingdom, Germany, and France (South Africa Tourism, 2021), a long-haul flight in excess of 10 hours is the norm.

While flying may be a more obvious source of CO_2 emissions, the impacts of terrestrial and oceanic mobilities such as cruise ships also need to be considered. Cruise ships are unique in that they serve as the mobility platform as well as the destination (Rink, 2020). The cruise industry contributes a disproportionate share of environmental degradation as argued by Klein (2010, p. 121) who warns that "on average cruiseships emit three times more carbon emissions than aircrafts, trains, and passenger ferries". Environmental pollutants from cruiseships not only include discharge of sewerage and 'grey' water that together induce algal blooms and inhibit the ocean's ability to absorb CO_2 (Klein, 2010) but also airborne pollutants from onboard incinerators and engines that burn extremely toxic and high sulphur content bunker fuel. Within the South African tourism sector, the planned growth of the cruise tourism threatens to bring additional environmental impacts. Much of that growth emphasises the development of new cruise tourists through 'cruises-to-nowhere' that sail into international waters while returning to the same port of departure (Rink, 2020), thus concentrating impacts within a contained geographical area spanning oceanic and terrestrial environments.

From the localised perspective of the destination, how tourists choose to move within it – known with the literature as 'modal choice' – is an under-explored question within tourism scholarship that requires further exploration (Baffi et al., 2020; Hannam et al., 2014). In South Africa, there is a disproportionate use of private automobiles by tourists relative to other tourism destination regions such as Europe and Asia where inter-city travel is made safe, easy, and reliable using long-distance rail services, while intra-urban travel can be fulfilled by public transportation services such as bus and metro systems. In a country where concerns over safety and security are paramount for visitors (Baffi et al., 2020; Donaldson & Ferreira, 2009), reliance on private vehicles increases automobile use in cities already choked with vehicular traffic. Modal choice for tourists is thus influenced by the range of options available at the destination and the particular contexts in which those options are situated. As Hannam et al. (2014) argue, tourists may feel restricted through the use of public transportation, while losing the sense of adventure and freedom that may come with travelling by automobile. While some have lauded the efforts of bus rapid transit systems in pulling tourists towards more public and sustainable modal choices (Baffi et al., 2020), the challenges of perception and the desire for freedom have an effect on mobility. Looking ahead, there is hope that more sustainable mobility options will be embraced by tourism planners and tourists themselves, including non-motorised transportation options as well as increased uptake of public transportation services (Hall et al., 2017).

Despite the negative aspects of tourism mobilities, they are vital for the growth and development of tourism generally, but especially in Africa where developing tourism's development potential is dependent on advances in aviation technology, allowing for more cost-efficient air access to (and within) the continent (Sarmento & Rink, 2016). Yet another factor in the 'wicked' problematic of climate change, a delicate balance must be struck between tourist mobility on the one hand and mitigation on the other.

Fundamental to the role of air travel in tourism-related GHG emissions has been the change in trends within the tourism sector over recent decades to more frequent but shorter trips and travel to more distant locations (Gössling & Peeters, 2007). Higham et al. (2016) refer to this behaviour as binge flying and air travel addiction. Both very short- and long-distance air travel is problematic. There is a large increase in the rate of emissions at take-off and landing (Becken, 2002; McKercher et al., 2010), but long-distance travel results in a significant accumulation of emissions across the journey. Long-distance travel is estimated to account for less than 3% of all trips, but is responsible for 70% of emissions relating to tourism (Rico et al., 2019). The GHG emissions related to the flight are seldom proportional to the time spent at the destination. An average holiday trip lasts for 4.15 days (Rico et al., 2019); yet, a tourist travelling from Australia to New Zealand would need to stay at their destination for >92 days for the GHG emissions related to the flight to be considered sustainable (Becken, 2002). Given the impact of air travel, a change in behaviour to travel less often and stay longer at destinations would be more appropriate (McKercher et al., 2010).

The greatest change to air travel patterns and greatest reduction in global GHG emissions since World War II emerged due to the COVID-19 pandemic rather than as a result of concern for the environment (Nguyen et al., 2021). On 11 March 2020, the day that the global pandemic was declared, Flightradar24 recorded 102,116 flights (Dube et al., 2021a); by the end of March 2020, there had been an 85% reduction in the number of flights compared to March 2019 and an 8% reduction in global emissions (Nguyen et al., 2021). By 26 April 2020, the minimum recorded number of flights on Flightradar24 was 24,049 (Dube et al., 2021b). However, these changes were short-lived. By mid-June 2020, the number of flights started increasing (Dube et al., 2021b), and air travel is once again rapidly increasing (Richie et al., 2021).

In southern Africa, air travel is considered to be of regional importance, particularly in attracting and transporting international tourists (DEAT, 2007; Turton & Matambira, 1996). The only study to date estimating the GHG emissions from aviation in southern Africa is for Victoria Falls, Zimbabwe (Dube & Nhamo, 2019b). It is estimated that 92,009,256 CO_2e kg is produced, of which 72,849,947 CO_2e kg relates to flights from South Africa (Dube & Nhamo, 2019b). In an attempt to reduce costs relating to fuel, which, in turn, reduced the GHG emissions, some of the airlines frequenting Victoria Falls are using the continuous descent approach, which has contributed to a 1.4% reduction in emissions (Dube & Nhamo, 2019b). Similar emissions reductions through fuel efficiency are reported globally, but as Higham et al. (2016) argue, the demand for travel far exceeds improvements in fuel efficiency. Moreover, as an airport related to a specific attraction, rather than one based in a large city, this is of course only a small component of the total carbon footprint of air travel in southern Africa. A large network of short- to medium-distance flights of 45 minutes to 4 hours connect the cities across southern Africa (DEAT, 2007) through national carriers (Figure 8.1). These are augmented by a large network of private commercial airlines and charter flights (Dube & Nhamo, 2019b).

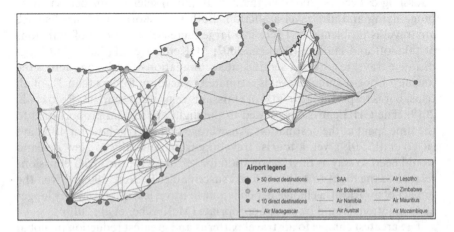

Figure 8.1 Airport traffic and national carrier air routes in southern Africa.

Table 8.1 Mode of transport into and out of South Africa, per category of visitor in January 2020

	Air (%)	Road (%)	Sea (%)
Total	• 28.5	• 70.2	• 1.3
South African	• 38.6	• 59.1	• 3.3
Foreign	• 19.4	• 79.8	• 0.8
Day Visitors	• 4.7	• 95	• 0.3
Tourists[a]	• 24.9	• 74.4	• 0.4
Overseas Tourists[b]	• 91	• 7.5	• 1.5
SADC Tourists	• 4.7	• 95.2	• <0.01
Other African Tourists	• 91.7	• 8.3	• <0.01

a Tourist defined as a person who spends at least one night in South Africa comprises 73.3% of border crossings in January 2020.
b Comprising Europe (65.1%), North America (13%), Asia (10.6%), Central and South America (5.2%), Australasia (4%), and Middle East (2.2%) in January 2020.
Data from Stats SA (2020).

In southern Africa, private cars are also extensively used, both in travelling to destinations, and in travel at destinations. A total of 3,930,440 travellers passed through South African ports of entry and exit in January 2020 (StatsSA, 2020), the majority of whom travelled by road (Table 8.1). While the majority of overseas tourists (91%) and those from outside of the SADC region (91.7%) travelled through ports of entry via air, 70.2% of all border crossings occurred by road (Table 8.1). Some argue that the predominance of road-based travel in the region is beneficial (Chen et al., 2018). First, it is easier to switch to lower or alternate fuel sources for motor vehicles than for aircraft (Chen et al., 2018). Second, a small car with two-person occupancy emits 60% less GHGs than an aircraft, per passenger kilometer (Gössling & Peeters, 2007). Furthermore, as much of the emissions from aircraft are released at cruising altitudes of 10 km–12 km in the lower stratosphere, the impacts on radiative forcing through clouds and contrails would increase the impact of emissions by 1.9–5.1 times that of a car (Gössling & Peeters, 2007; Lenzen et al., 2018; Rico et al., 2019). A shift from air to overland travel, where possible, is therefore important, particularly for shorter journeys within or between countries (Scott et al., 2010; Steyn & Spencer, 2012). Ideally, this should involve changes to rail travel in the long run (Steyn & Spencer, 2012).

For some countries, such as China, private cars are becoming a primary source of tourism-related GHG emissions (Chen et al., 2018; Huang et al., 2017). An increase in self-driving tours, stimulated by improvements in living standards and in transport facilities, is resulting in significant congestion and, in turn, heightened emissions during peak tourist periods (Huang et al., 2017; Jin et al., 2018). Moreover, at destinations where air travel is common, such as Victoria Falls, there is an unquantified but extensive use of private cars following a tourist's flight, traversing small distances between accommodation and the waterfall (Dube & Nhamo, 2021). Tour operators at the Victoria Falls additionally make extensive use of large, fuel-inefficient 4 × 4 vehicles, while accommodation establishments are reported to have large fleets of vehicles to shuttle tourists and for daily operations (Dube & Nhamo, 2021). For transport between attractions at destinations, it is therefore important to make public transport more appealing to tourists both through the provision of quality services and the development of ticketing conducive to tourists (Chen et al., 2018; Jin et al., 2018). At an extreme, this may involve banning private cars within tourist spaces (Chishti et al., 2018), but a less hard-line approach would involve a move to more fuel-efficient cars, particularly for rental car fleets (Dube & Nhamo, 2019b, 2021; Jin et al., 2018; Scott et al., 2010). Indeed, policies relating to GHG emissions from private transport are aimed at everyday travel, and tourist travel is longer and more energy intensive (Higham et al., 2016).

Surplus and excess while on holiday – where usual practice disappears

While transport undoubtedly represents the greatest proportion of the role of tourism in GHG emissions and, in turn, contributing to climate change, the role of tourist activities at their destination is not negligible (Scott et al., 2010). Tourist accommodation is estimated to represent 21% of the GHG emissions from tourism (Scott et al., 2010) and argued to be the largest contributor of GHG emissions at Victoria Falls (Dube & Nhamo, 2021). Tourist accommodation establishments serve as the primary supplier of the energy and water that tourists will use during their vacation and thus have the greatest potential to drive or to mitigate GHG emissions at destination scale (Hoogendoorn et al., 2015). It is sometimes argued that the water and energy used by tourists, whether directly, or through food, shopping, and ground transport, would have been used regardless of whether they were on vacation, and that only the additional use – a component considered 'additionality' – should be considered as relating to tourism (Shakouri et al., 2017). However, the use at a destination affects the carbon footprint of the destination, and so any consumption that would have occurred at home, or indeed which might continue to occur at home during their vacation, should therefore be considered a red herring (Lenzen et al., 2018; Shakouri et al., 2017). Moreover, on the whole, consumption is estimated to be far greater on vacation than during ordinary day-to-day life (Sun & Hsu, 2019).

In southern Africa, there is a considerable reliance on coal-derived electricity (Amusan & Olutola, 2017; Dube & Nhamo, 2021). As a result, electricity used in lighting, heating, cooling, and cooking in accommodation establishments contributes to GHG emissions (Dube & Nhamo, 2021; Hoogendoorn et al., 2015). The amount of energy used is highly spatially heterogenous. In warm regions such as KwaZulu-Natal Province of South

Africa, and Victoria Falls in Zimbabwe, air-conditioning and fans are used extensively, heightening the energy use compared to more temperate regions (Dube & Nhamo, 2021; Hoogendoorn et al., 2015). Where air-conditioning is retrofitted into existing rooms, rather than in accommodation designed with artificial airflow in mind, the energy use is heightened further (Dube & Nhamo, 2021). Older accommodation establishments are also often equipped with older appliances, which do not have energy ratings and which deteriorate in their energy efficiency over time (Mearns & Boschoff, 2017). In these instances, improvements in technology to improve energy efficiency, such as the installation of energy saving or LED lightbulbs as in the case of guesthouses in Mpumalanga (Machete et al., 2016), Gauteng and KwaZulu-Natal (Hoogendoorn et al., 2015), and Grootbos Private Nature Reserve (Dube & Nhamo, 2020e) in South Africa, or more extensive implementation of low-energy appliances and solar energy as for Hotel Verde Cape Town (Dube & Mearns, 2019), would reduce both tourism-related GHG emissions and local energy costs. However, the perceived high upfront costs of these changes often serve as a barrier, even where longer-term energy-related cost-savings are likely (Gabarda-Mallorquí et al., 2017).

Machete et al. (2015) argue that energy savings need to be matched against tourist comfort to ensure the economic success of accommodation establishments. Often electronic appliances, and the comfort they offer, serve to enhance the competitive advantage of a particular accommodation listing (Prinsloo, 2019). In the case of Airbnb in Cape Town, across 500 listings sampled, a large array of large-, medium-, and small-sized appliances are listed, many of which would not be considered necessary for a touristic visit to a major city (Table 8.2). Notably, despite on-going load-shedding in South Africa, only 24 of the 500 listings included warnings regarding excessive electricity use, some of which charged an additional levy for using large appliances, and some had prepaid electricity with a per-day allowance (Prinsloo, 2019).

Water use by tourists, whether at accommodation establishments, attractions, or restaurants does not directly lead to GHG emissions, but is rather a problem in worsening water shortages for both local residents and the tourism sector in arid, water-stressed regions, or in periods of drought (Sun & Hsu, 2019). Water is used by tourists in cleaning of their accommodation, for their personal hygiene, the removal of sewage, food production and preparation, and for scenery and recreation (Sun & Hsu, 2019; Zhang et al., 2017). Water use can also indirectly contribute to GHG emissions through the purification, supply, heating, and treatment of water (Baumber et al., 2021), with adaptations such as low-flow showerheads serving to reduce both the water and carbon footprint (Dube & Nhamo, 2021; Hoogendoorn et al., 2015; Prinsloo, 2019). Compact and vertical configurations that are common in mass tourism resorts are more water and energy efficient per tourist night than low-density tourist destinations, yet higher graded hotels use more water overall through supplying the additional features of gardens, swimming

Table 8.2 Appliances listed on Airbnb profiles for Cape Town City Bowl and surrounds, 2018

a) Size of appliance	Appliance	Number of listings (n = 500)	Likely frequency of use
Large	• Washing machine	• 385	• 1–2 times a week
	• Tumble dryer	• 185	• 1–2 times a week
	• Washer dryer	• 6	• 1–2 times a week
	• Dishwasher	• 85	• Daily, if self-catering
	• Jacuzzi/hot tub	• 31	• Daily (seasonal)
	• Stove and oven	• 165	• Daily, if self-catering
	• Double oven	• 3	• Infrequent
	• Oven	• 47	• Daily, if self-catering
	• Fridge/freezer	• 265	• Multiple times per day
	• Deep freeze	• 3	• 1–2 times a week
	• Air conditioner	• 137	• Daily (seasonal)
	• Heated floors	• 5	• Daily (seasonal)
	• Electric Geyser (Hotwater)	• 1	• Daily
Medium	• Bar fridge	b) 22	• Multiple times per day
	• Heater	• 263	• Daily (seasonal)
	• Television	• 329	• Daily
	• Microwave	• 250	• Daily
	• Electric blanket	• 6	• Daily (seasonal)
	• Computer	• 1	• Infrequent
	• Hot plate/stove	• 53	• Daily, if self-catering
	• Fan	• 43	• Daily (seasonal)
Small	• Electric kettle	• 271	• Multiple times per day
	• Food blender	• 15	• Infrequent
	• Toaster	• 193	• Daily
	• Coffee machine	• 89	• Daily
	• Juicer	• 3	• Infrequent
	• Icemaker	• 1	• Daily
	• Heated towel rail	• 5	• Daily (seasonal)
	• Hairdryer	• 305	• Daily
	• Iron	• 374	• 1–2 times a week
	• Steamer	• 1	• 1–2 times a week
	• Light dimmers	• 1	• Daily
	• Wifi/Internet router	• 465	• Daily

Data collected in collaboration with A Prinsloo.

pools, and spas (Gabarda-Mallorquí et al., 2017). Notably, despite the marketing surrounding greening efforts made by large chains (Hoogendoorn et al., 2015), independent hotels are found to use less water than chains, perhaps because the cost-savings associated with reduced water consumption are felt more acutely by smaller enterprises (Gabarda-Mallorquí et al., 2017).

While the water footprint of sewage, shopping, and food in the tourism sector are often discounted or underestimated (Sun & Hsu, 2019; Zhang et al., 2017), the greatest source of water relates to golf tourism (Scott et al., 2018). In 2004, an estimated 9.5 billion liters of water was used per day in irrigating golf courses internationally (Scott et al., 2018). The use of water at golf courses is calculated to be the equivalent, per tourist day, of double that of a five-star room, which at least a proportion of golf tourists will also occupy (Graciano et al., 2020). Indeed, Scott et al. (2018) raise concern that studies on water use in tourism often focus exclusively on accommodation, and the much larger use from golf courses is ignored (Scott et al., 2018). Water use in golf tourism doubles during anomalously warm and dry seasons, heightening the water stress experienced by local communities (Graciano et al., 2020; Scott et al., 2018). In Mauritius, in part due to the extensive golf courses, 10% of national water is used by the tourism sector (Sun & Hsu, 2019).

BOX 8.2 AIRBNB TOURISM ACCOMMODATION DURING THE CAPE TOWN DROUGHT

Ariel Prinsloo

Traditional tourism accommodation providers such as hotels and guest houses have come to compete with the likes of more informal tourism accommodation comprising the sharing economy such as Airbnb. The online platform has become more popular due to its diverse accommodation-type offerings and claims to provide a more authentic experience for tourists (Visser et al., 2017). Accommodation types can be categorised according to shared or private rooms in which a host and guest share a space, and the rental of entire homes or apartments, where there is often limited interaction with the host. The nature of the platform allows for the potential to reduce resources through shared space; however, with its increased popularity and the resource demands of tourists, this may either encourage or impede sustainable resource use (Martin, 2016; Richardson, 2015). This is of particular interest during drought events as seen in the 'Day Zero' drought in Cape Town, South Africa.

The City of Cape Town, located in the southwest point of the Western Cape in Southern Africa, is a tourism hotspot, welcoming approximately two million tourists annually (Cathala et al., 2018). This region experienced a three-year drought from 2015 to 2018, which at the time was forecast to lead to a 'Day Zero' in which the supply of water to the city would run out and residents would be forced to queue for daily rations of water (Dube et al., 2020). Cape Town is characterised by a Mediterranean climate with warm dry summers and mild wet winters. The 'Day Zero' drought was driven by a poleward shift of the westerly moisture corridor, resulting in three consecutive years of reduced winter rainfall (Dube et al., 2020; Sousa et al., 2018). With dam supply levels critically low during this period, the City of Cape Town put into effect water restrictions where should usage exceed the permitted amount, residents could be heavily fined. The City also aimed to reduce water use through campaigns such as *Save like a local* and *Think Water*, where standard practices such as two-minute showers and minimal flushing were promoted (Brick & Visser, 2018).

In the midst of the drought, Airbnb was launched in Cape Town in July 2015 with 17,600 listings as of 2017 (Visser et al., 2017). As compared to formal tourism accommodation such as hotels, Airbnb providers were restricted by residential water restrictions, only receiving warnings when water use exceeded the permitted level (Prinsloo, 2019). Further, the identity of Airbnb properties by municipalities is not known and thus cannot be causally linked to tourism water use, relying on the host as a Capetonian resident to implement restriction levels. On the one hand, through shared properties, the close proximity to guests would allow hosts to be able to easily monitor guest movement, intercepting when water usage was high. They were also able to implement additional water restriction methods, for example, through intimate conversations, turning off of toilet water supply and limiting laundry services. In addition, the decisions of water use lay directly with the host, with some hosts not bathing to allow guests to use their allotted water amount and others banning guests from using their personally collected spring water (Prinsloo, 2019). On the other hand, rental of entire homes or apartments allows hosts to be able to take a hands-off approach, with very few hosts actually advertising about the drought in their online profiles and some believing that guests had no substantial impact on their property water use at all. In addition, the increasing use of Airbnb management firms and hosts managing on behalf of other host properties, who were revealed to be predominately motivated by income or supplemental income, did little to enforce water restrictions besides signage and brief mention of the drought on their online profiles (Palgan et al., 2017). Furthermore, many private apartments rented were also sectional titles, whereby decisions and monitoring around water saving are controlled by body corporates and thus difficult to enforce.

With the likelihood of increased drought events under climate change, this example reveals brief insight into the potential of tourism accommodation providers, although informal, to determine the extent to which the impact of drought is worsened or averted by tourism resource use behaviour (Marshall et al., 2011).

Activities at tourist attractions are also becoming a key contributor to the GHG emissions within the sector. In China, between 2000 and 2018, the proportional contribution of accommodation declined from 19% to 7%, while that of tourist activities increased from 10% to 18% (Chen et al., 2018). For

Victoria Falls in Zimbabwe, 15 helicopters and 20 cruise boats are in operation, with 15-minute helicopter trips taking place throughout the day (Dube & Nhamo, 2021). At a broader scale, tourist activities also contribute to heightened plastic waste along coastlines both from terrestrial activity and boat trips (Eagle et al., 2016), affecting the destination image of the beach and the quality of life of residents (Wilson & Verlis, 2017). Plastic waste on beaches has also been implicated in local daytime warming of beaches, with beach soil temperatures increasing by 2.45°C (Lavers et al., 2021). Waste generation – both in terms of litter and managed waste at accommodation establishments and attractions – further contributes to GHG emissions through energy use in the collection and processing of waste and methane emitted from the decomposing waste (Baumber et al., 2021). Air pollution, often including but not limited GHGs, is likewise released through tourist activities at accommodation establishments tourist activities at accommodation establishments and transport, compromising the air quality of the tourist destination for up to two months for each month of pollution (Zhang et al., 2020).

Calculating the carbon footprint of tourism

Given the wide-ranging sources of tourism-related GHG emissions, effective mitigation would benefit from the accurate and effective quantification of the emissions throughout the tourism life cycle (Lenzen et al., 2018). At present, a range of calculation techniques, boundary definitions, and input data exist, with the methodology and level of complexity having a significant impact on the results obtained and the ability for comparison (Meng et al., 2016). This is largely due to the fragmentation of tourism sectors and the interconnectedness of emission-intensive services used by tourists and residents (Akadiri et al., 2019b; Balsalobre-Lorente et al., 2020; Nowak et al., 2010). Transportation, entertainment, catering, and natural resources are increasingly used by both tourists and residents, as tourists are increasingly self-booking their trips and are not constrained to packaged experiences (Nowak et al., 2010). It is therefore imperative to recognise the economy of a destination as an interlinked network of sectors, each of which perform transactions with one another and share in their carbon emissions (Sun et al., 2020). It is also important to delineate boundaries as to which components, and the proportional emission contributions thereof, are included in tourism carbon footprints (Sun et al., 2020).

The greatest challenges relate to the inclusion of international air travel in local and/or sector-specific GHG emission inventories (Becken, 2002). International air travel is seldom included in national inventories (Becken, 2002), and there are no binding agreements for international aviation in the Paris Agreement (Lenzen et al., 2018; Richie et al., 2021). There is also ambiguity as to how international aviation would be included in national inventories; Becken (2002) puts forward three options of (1) accounting for fuel burned within 200 km of the economic zone, (2) fuel purchased in the country, or (3)

50% of the fuel consumed between origin or destination. Spurious results can also be obtained when attributing the full proportion of air travel to a destination and in particular to a specific activity. Using life-cycle calculations, Sun and Drakeman (2020) calculate that cellar-door sales are the highest contributor to GHG emissions of the wine production process at seven times that of packaging. This is because, for the cellar-door visitor, their full GHG emission of travel from home to wine farm are calculated, including the GHG emissions of their flight to the destination country, without distributing the GHG footprint across all activities or attractions the tourist partakes in.

A large toolbox of analytical tests and models can be applied in calculating the carbon footprint of the tourism sector including Granger Causality Analysis (see Akadiri et al., 2020), Multi-Variate Regression Models (see Azam et al., 2018), Factor Decomposition Analysis (see Chen et al., 2018; Tang et al., 2017), Life Cycle Assessments (see Jin et al., 2018; Sun & Drakeman, 2020), and Ecological Footprint Analyses (see Hoogendoorn et al., 2015). These can be computed either as a top-down or bottom-up approach, depending on the type of data available (Sun & Drakeman, 2020). These could include flight data such as the IATA schedule reference analyser used in the ICAD carbon emissions calculator for Zimbabwe (Dube & Nhamo, 2019b) or toll-gate data used in calculating emissions from private cars (Jin et al., 2018). More recent approaches, aiming to provide a more comprehensive model and replace traditional bottom-up or top-down models, include the Environmentally Extended Input-Output Model (Sun et al., 2020). These models should consider the life cycle or supply chain of emissions and include GHGs emitted during tourist activities and those embodied in services and commodities purchased by tourists (Lenzen et al., 2018). The models can also account the carbon emissions by the country of residence of the tourist or by the destination where the GHGs are emitted (Lenzen et al., 2018). To date, carbon footprint calculations in southern Africa have only been applied to Victoria Falls (Dube & Nhamo, 2019b, 2021) and ecological footprint calculations to a selection of guesthouses in Gauteng and KwaZulu-Natal Provinces in South Africa (Hoogendoorn et al., 2015).

Given the wide range of possible methods, and the role of methodology in determining the output value, it is important that any carbon footprint calculation clearly discloses all assumptions made, limitations in data, and definitions of boundaries that apply (Sun et al., 2020). Meng et al. (2016) argue for the urgent need for a unified measurement framework for tourism carbon emissions. By contrast, Becken (2017) argues that low-carbon is more appropriate and achievable than carbon-neutral and would allow for a shift from quantifying emissions to the promotion of low-carbon practices and values (Baumber et al., 2021).

Greening efforts in the southern African tourism sector

The global tourism sector had self-imposed GHG emission reduction goals of 25%–30% by 2020 and 50% by 2050 (Scott et al., 2010). Even with the significant reduction in flights and tourist activities in the second quarter of 2020 due to COVID-19, the goals for 2020 were not met, and those for 2035 appear unlikely to be attained (Nguyen et al., 2021; Richie et al., 2021). Strategies to mitigate the impact of tourism on climate change include the reduction of emissions through improvements in energy efficiency and changes to renewable power and biosequestration (Baumber et al. 2021; Scott et al., 2016c; Steyn & Spencer, 2012). These can broadly be classified under the umbrella of greening – defined either as gradual processes driven by small changes that provide returns in cost cutting or more strictly as deliberate actions to mitigate the harms to the environment caused by tourism (Hoogendoorn et al., 2015). There is some evidence of greening in southern Africa. In KwaZulu-Natal, Gauteng, and Mpumalanga, tourism accommodation establishments are introducing low-energy lights, solar power, and low-flow showerheads (Hoogendoorn et al., 2015; Machete et al., 2015), while during the 'Day Zero' drought in Cape Town, the tourism sector is argued to have reduced its water usage by 50% (Dube et al., 2020). However, perceptions of aesthetics and preference for water- and energy-intensive goods and services remain (Machete et al., 2015; Scott et al., 2018; Sun & Hsu, 2019), and many operators believe that individual change is irrelevant in the face of uncertainty (Eagle et al., 2016; Gössling & Peeters, 2007). This arguably requires stronger role of policy and regulation in ensuring engagement and compliance in emission reductions.

Greening, low-carbon or carbon-neutral accreditation, rating, and certification schemes provide one avenue to achieve this (Baumber et al., 2021; Chen, 2019). Globally, more than 100 eco-certification schemes have emerged over the past three decades (Baumber et al., 2021), arguably allowing for the diagnostic assessment of GHG emissions and providing incentives to operations which successfully attain their target reductions (Gabarda-Mallorquí et al., 2017). However, there is debate as to whether rating or accreditation provides a tangible competitive advantage, and whether this leads to more sustainable tourist and operator behaviour (Baumber et al., 2021). Indeed, it has been found that there is little difference in the GHG emissions and broader sustainability of certified and uncertified groups when maximising revenue is a greater priority to the organisation than the greening process (Chen, 2019). Key attributes of an effective rating scheme are therefore the incentivisation of emission reductions, the encouragement of collaboration between sectors, low costs in implementation, and an accreditation process which is deemed to be fair, rigorous, and objective (Baumber et al., 2021). In southern Africa, there is currently little uptake for these schemes. Only 20% of accommodation establishments interviewed at Victoria Falls were accredited by Green Tourism, an initiative operated by the Zimbabwean Department of Environmental Affairs and Zimbabwean Tourism (Dube & Nhamo, 2021). Likewise, the majority of guesthouses in Gauteng and KwaZulu-Natal Provinces in South Africa which were interviewed on greening practices were not green-graded (Hoogendoorn et al., 2015).

A second approach is that of either compulsory or voluntary carbon offsets, at cost to either the tourist or the operator. To date, these have predominantly involved air travel. Voluntary carbon offset programmes that devolve responsibility onto the consumer with an impact on their cost to fly have been met with considerable scepticism both in terms of efficacy and the apportionment of responsibility (Higham et al. 2016; Richie et al., 2021). Only 1%–10% of passengers have been found to purchase voluntary carbon offsets, and those who do fall into a narrow category of travellers who are under the age of 30, work full time, and, importantly, are not frequent fliers (Richie et al., 2021). Offsetting therefore needs to be driven by the tourism sector or airline itself (Steyn & Spencer, 2012). The International Civil Aviation Organisation introduced a carbon offset and reduction scheme for international aviation in 2016, which proposes compulsory carbon offsets from 2027 for all countries except Small Island States (Richie et al., 2021). In South Africa, the organisations SATSA and SA Tourism have launched a carbon offsetting initiative, planning to plant 5000 spekboom, a species of high carbon sequestering succulents indigenous to the Eastern Cape, as a means to reduce the carbon footprint of inbound flights (SATSA, 2019). However, as Gössling (2000) calculated at the time that 30,000 km^2 of land needs to be afforested annually to offset global emissions from tourist air travel, this is unlikely to make a significant contribution.

Social marketing is argued to show more promise in influencing human behaviour (Eagle et al., 2016; Higham et al., 2016), but is probably insufficient when pitted against commercial marketing that encourages international flights, vacations to distant locations, and high-energy consumption (Higham et al., 2016). Taxes or tax rebates may therefore be the most effective approach (Sun & Hsu, 2019; Zhang & Zhang, 2018). Carbon taxes on aviation and tourism operations are perceived to be able to effectively reduce tourism-related carbon emissions, but at significant economic cost to the tourism industry, reducing the tourism contribution to Gross Domestic Product (GDP) and threatening jobs related to tourism (Zhang & Zhang, 2018). If these taxes were recycled back into the tourism sector, some of the harms from those taxes may be minimised, but the damage induced by a reduction in the number of incoming tourists would be unavoidable (Zhang & Zhang, 2018). As tourism is the only sector in South Africa that managed to increase jobs while increasing its contribution to GDP, this could be detrimental to the economy at large (DEAT, 2007). A tax credit to operators who meet emission reduction targets may be more appropriate (Sun & Hsu, 2019), or fare rebates to tourists who minimise their water or energy use, as is successfully applied at Hotel Verde in Cape Town, South Africa (Dube & Mearns, 2019). At present, in southern Africa, as for most regions globally (Gössling & Peeters, 2007), private cars, shuttles, and buses are currently taxed on vehicle fuel, tollgates, and value-added tax (VAT). Aviation fuel and emissions are not currently taxed (Gössling & Peeters, 2007), but the challenges of discouraging tourist arrivals in tourism-dependent economies would apply (Zhang & Zhang, 2018). Alternate approaches include adjusting airfare costs to better align with the true cost of the flight to discourage inefficient flights with distant layovers, emissions trading, and providing viable alternatives to air travel (Becken, 2002).

Reducing the impact of the tourism sector on GHG emissions and climate change requires a combination of technological solutions, behavioural change among tourists and tourist operators, and local strategies at and between destinations (Baumber et al., 2021). In light of this, it is concerning that empirical research has found that regular international tourists are the most aware of climate change, and the least likely to change their behaviour, while those who travel less frequently are most willing to reduce their emission profiles further (McKercher et al. (2010). A combination of approaches targeting both tourists and operators, through voluntary and compulsory mechanisms, may therefore be most prudent.

Conclusion

GHG emissions from the tourism sector are increasing as a result of a rapidly growing number of people taking part in domestic and international tourism, and those individual tourists travelling to more distant locations, more frequently, for shorter periods of time. While the sector currently comprises a low proportion of global emissions, when compared to industries such as manufacturing, it is important to monitor changes in emissions and to work towards a continued transition to low-energy technologies. The IPCC Special Report on Global Warming of 1.5°C (Hoegh-Guldberg et al., 2018, p. 5) provides some hope, arguing that "reaching and sustaining net-zero global anthropogenic CO_2 emissions and declining net non-CO_2 radiative forcing would halt anthropogenic global warming on multi-decadal timescales"; the more recent Sixth Assessment Report of the IPCC (2021) argues for the urgency of reaching net-zero emissions, and the likelihood that any stabilisation in warming trends will only be experienced in the second half of the century. Tourism has the potential to not only achieve a transition to low-carbon within the sector but also inspire a more widespread carbon transition in both inter-connected and distal sectors through redistributing wealth, addressing socio-economic inequalities, contributing to economic growth and development, and diversifying the nation's economy, while ensuring environmental sustainability. The tourism sector could also potentially provide good starting point for accreditation, rating, and certification schemes that can be expanded out to other sectors once buy-in is secured within the tourism sector. At present, effective approaches within the tourism sector involve a combination of social marketing, combined at times with differentiation of destinations and operations via grading schemes, voluntary offsets, emissions trading, and taxes, each applied to specific components of the sector, and with varying levels of influence.

9 Governance and policy needs in tourism and climate change relations

Introduction

Over the past decade, governance has become one of the keywords in tourism destination management studies (Bramwell et al., 2017; Bramwell & Lane, 2011; Wray, 2015), with close links to sustainability and climate change policies and actions (Hall, 2011; Wyss et al., 2014; Zeppel, 2012). This is also the case in southern African tourism scholarship (Cornelissen, 2017; Dube et al., 2020; Dube & Nhamo, 2019a), with increasing calls for climate change governance and supporting policy frameworks for sustainable and climate-concerned tourism development (see Hambira & Saarinen, 2015; Pandy & Rogerson, 2021; Rogerson, 2016). In general, the United Nations Development Programme (UNDP, 1997) has stated that governance involves institutions, systems, processes, and relationships through which various stakeholders can articulate their interests, goals, and negotiate their different positions. In the context of tourism, the World Tourism Organisation (UNWTO, 2008, pp. 31–32) has defined tourism governance as the

> ...process of managing tourist destinations through synergistic and coordinated efforts by governments, at distinct levels and in different capacities; civil society living in the inbound tourism communities; and the business sector connected with the operation of the tourism system.

In southern Africa, governance discussions are often related to the idea of 'good governance' (Nunkoo, 2017). There are many different aspects, levels, and indicators concerning governance and how 'good' it is (see Smith, 2007; Sundaram & Chowdhury, 2012); the extensive discussions and debates about this are beyond the scope of this chapter. In short, however, nobody wants to advocate bad governance for the sake of development. Thus, concerning development and sustainability, we need to ask: What is good governance for whom and by whom? This is crucial especially when foreign or external actors and donors are actively involved with governing (tourism) development (or growth) and its outcomes in local contexts. Generally, good governance refers to transparency, accountability, uncorrupted administration, participation, and efficient institutions (Gisselquist, 2012).

DOI: 10.4324/9781003102618-9

In tourism, we can interpret good or better governance as a framework that is advocated for the industry and its key stakeholders to create socioeconomic development and well-being for local communities and the environment (Siakwah et al., 2020). This connects good governance discussions, focusing on the needs of sustainable development in tourism (Butler, 1999; Saarinen & Gill, 2019; Wan, 2012). Indeed, as indicated by Wray (2015), good governance is necessary for tourism development and for destinations to achieve sustainable outcomes (see Bramwell & Lane, 2011), and the idea of sustainable tourism has been highly influential in how governments and practitioners have formulated various tourism policies in southern Africa (Nunkoo, 2017).

The connections between tourism development, good governance, and sustainability are often complex and sometimes conflicting (Sharpley, 2000). This is especially the case with the climate change process, which is a much more contested issue than many others in the sustainable tourism discussions (Bramwell et al., 2017). A primary reason is that tourism development and its good governance often mean growth thinking in practice (see Christie et al., 2013; Saarinen, 2021b). This can be highly problematic in relation to the politics of climate change (Hall, 2009; Gössling et al., 2021) and for the sustainability of tourism (Saarinen, 2021a). Thus, while the idea of sustainable tourism was a major social innovation for the sector in the 1990s (Hall, 2011), there are urgently evolving needs for policy and governance innovations in sustainable tourism development (Bramwell & Lane, 2011). This is the case for both climate change (Bramwell et al., 2017; Gössling & Scott, 2018) and the United Nations SDGs (Saarinen, 2020; Scheyvens, 2018; Siakwah et al., 2019) (Box 9.1). These two aspects and their integration in destination governance are critical for the future of tourism development in the southern African region.

BOX 9.1 SUSTAINABLE DEVELOPMENT GOALS AND THE HOSPITALITY SECTOR IN ZIMBABWE

Ngoni Courage Shereni and Kaitano Dube

The year 2015 is a watershed moment in the sustainability discourse which saw the United Nations (2015a) introducing the 17 SDGs and the 169 related targets. Even though there is a belief that tourism can contribute to all the 17 SDGs, the Agenda 2030 document makes direct reference to three of them: SDG 8 on inclusive and sustainable economic growth; SDG 12 on sustainable consumption and production; and SDG 14 on life below water. There is overwhelming evidence that the hospitality sector, as a major part of the tourism sector, can make significant contributions to the achievement of the SDGs (Dube & Nhamo, 2020a; Liburd et al., 2020; Saarinen, 2020).

Studies focusing on the adoption of SDGs in the tourism industry in Zimbabwe are scant, although this field of research is growing (Siakwah et al., 2019). A number of existing studies in the Zimbabwean context have focused on the environmental sustainability dimension (Chigora et al., 2019; Dube & Nhamo, 2020a; Njerekai, 2019; Zengeni et al., 2013). Njerekai (2019) notes that the adoption of sustainable practices by the hospitality sector in Zimbabwe is low. More so, the SDGs have not been fully adopted by the hospitality sector in Zimbabwe due to a number of challenges. These challenges include the poor performance of the hospitality sector with low occupancy rates averaging 50% per annum, a lack of financial resources, the high cost of installing sustainability technologies, and a lack of knowledge (Dube & Nhamo, 2020a; Mbasera et al., 2016).

While Njekerai (2019) argues that the hospitality industry in Zimbabwe has been lagging in embracing sustainability, other tourism scholars have indicated that some practices of sustainability have been embedded in the tourism and hospitality sector (Dube & Nhamo, 2020b). In the Victoria Falls, for example, the hospitality sector is making efforts to address the greenhouse gas (GHG) emissions as part of efforts aimed at addressing climate change (SDG 13) and reducing waste (SDG 12) through Green Tourism accreditation (Dube & Nhamo, 2020b). In 2016, approximately 20% of accommodation establishments had been accredited under various categories of Green Tourism in Victoria Falls in line with the ethos of Zimbabwe National Parks of business "in harmony with nature" (Dube & Nhamo, 2020b, p. 129). The use of local resources that aid in climate change mitigation and adaption is a common feature of tourism enterprises in the Victoria Falls, where thatching and growing of trees fall under the purview of SDGs 13 and 15.

An analysis of the website content of four major hospitality chains in Zimbabwe (African Sun Limited, Rainbow Tourism Group, Cresta Hospitality and Africa Albida Tourism) reveals that there is disclosure of SDGs and sustainability issues, primarily located in their annual reports. Smaller establishments do not give much attention to the SDGs because of the nature of their operations and a lack of information. It has been observed that the adoption of sustainable practice in general in Zimbabwe is driven by voluntary practices. Thus, there is no blueprint to guide and compel hospitality establishments to implement the SDGs (Mbasera et al., 2016).

There is a notable increase in the adoption of SDGs in the hospitality sector in Zimbabwe. In the absence of a strong legal framework compelling establishments to adopt sustainable practices, the adoption of SDGs is voluntary. Environmental practices such as installation of solar technologies, waste recycling, water-saving initiatives, and afforestation among others are the most popular practices.

Governing sustainable tourism and innovation policies

Governance is multi-dimensional (Rhodes, 1996, 1997). Ruhanen et al. (2010), for example, analysed 53 governance studies in political science and management literature, from which they identified 40 different dimensions of governance. These dimensions include accountability, transparency, involvement, structure, effectiveness, power, efficiency, and decentralisation, for example, which have been applied to destination governance and sustainable tourism studies (see Saarinen & Gill, 2019). In sustainable tourism literature, governance refers to the process of collaboration and collective action between various public and private sector stakeholders in a multi-scalar policy environment towards sustainability (Bramwell & Lane, 2011; Gill & Williams, 2011). In a climate change context, sustainability governance aims at strengthening destination resilience and adaptive capacity; processes that have faced challenges in implementation in the past (Hall, 2011; Hall et al., 2018; Saarinen & Gill, 2019). Related to these challenges, Bramwell (2011) has identified two interwoven reasons: multi-scalarism and coordination.

The former refers to the fact that destination governance involves cross-sectoral relations and diverse policies at various levels of planning, development, and management (Hall, 2000; Saarinen et al., 2017). These levels operate in local, regional, national, and supra-national scales that may have a variety of tourism- and non-tourism-specific policies. Both tourism- and non-tourism-specific policy types can have implications affecting the sector and its development prospects, and together with other policies operating at different administrative levels, destination governance becomes a highly complex structure to manage efficiently. With respect to the latter coordination challenge, the multi-scalar nature of tourism and related policies creates a highly complex and confusing arena of institutional arrangements for collaboration and well-coordinated actions (Hall, 2011; Saarinen & Gill, 2019; van der Duim et al., 2011). This has been noted as one of the key problems in the governance of tourism development in relation to climate change policies and actions (see Bramwell et al., 2017; Gössling & Scott, 2018; Wyss et al., 2014; Zeppel, 2012). In addition, the tourism sector itself is a highly fragmented industry involving a diverse set of operators and stakeholders ranging from the public to private sectors and from hosts to guests (Gill & Williams, 2011). Different stakeholders have unique interests, priorities, values, and varying capacities related to tourism development and its (good) governance. In this respect, however, locals' capacity to influence the course of tourism development has been noted as problematic in southern Africa (see Fairer-Wessels, 2017; Musavengane et al., 2020; Saarinen et al., 2009).

Despite these past implementation challenges, many international development organisations and national governments in southern Africa consider tourism as a highly prospective tool for regional development and community empowerment (Rogerson & Saarinen, 2018). Recently, these prospects have been linked with the United Nations SDGs (United Nations, 2015a) and ideas on how tourism could contribute to the UN Agenda 2030

Table 9.1 The targets and specific indicators for the UN Sustainable Development Goal 13: Take urgent action to combat climate change and its impacts

Targets	Indicators
3.1 Strengthen resilience and adaptive capacity to climate-related hazards and natural disasters in all countries	13.1.1: Number of deaths, missing persons, and directly affected persons attributed to disasters per 100,000 population 13.1.2: Number of countries that adopt and implement national disaster risk reduction strategies in line with the Sendai Framework for Disaster Risk Reduction 2015–2030 3.1.3: Proportion of local governments that adopt and implement local disaster risk reduction strategies in line with national disaster risk reduction strategies
13.2 Integrate climate change measures into national policies, strategies, and planning	13.2.1: Number of countries with nationally determined contributions, long-term strategies, national adaptation plans, and adaptation communications, as reported to the secretariat of the United Nations Framework Convention on Climate Change Total greenhouse gas emissions per year
13.3 Improve education, awareness-raising, and human and institutional capacity on climate change mitigation, adaptation, impact reduction, and early warning	13.3.1: Extent to which (i) global citizenship education and (ii) education for sustainable development are mainstreamed in (a) national education policies; (b) curricula; (c) teacher education; and (d) student assessment
13.A Implement the commitment undertaken by developed-country parties to the United Nations Framework Convention on Climate Change to a goal of mobilising jointly $100 billion annually by 2020 from all sources to address the needs of developing countries in the context of meaningful mitigation actions and transparency on implementation and fully operationalise the Green Climate Fund through its capitalisation as soon as possible	13.a.1: Amounts provided and mobilised in United States dollars per year in relation to the continued existing collective mobilisation goal of the $100 billion commitment through to 2025
13.B Promote mechanisms for raising capacity for effective climate change-related planning and management in least developed countries and small island developing States, including focusing on women, youth, and local and marginalised communities	13.b.1: Number of least developed countries and small island developing States with nationally determined contributions, long-term strategies, national adaptation plans and adaptation communications, as reported to the secretariat of the United Nations Framework Convention on Climate Change

Source: United Nations (2015a).

(Hughes & Scheyvens, 2016; Saarinen, 2020). One of the key values of SDGs is 'leave no one behind', which aims to represent the unequivocal commitment of all UN member states to eradicate poverty, end discrimination and exclusion, and reduce the inequalities and vulnerabilities associated with global economic development. This principle should be central to sustainable tourism development, with a focus on the well-being of people and their living environments.

In general, the SDGs represent potential social and policy innovations defining the agenda for global development. In total, there are 17 goals and 169 specific targets set to transform the world by the year 2030 (United Nations, 2015a). There is a specific SDG for sustainable innovations (SDG9), and while all the SDGs are interrelated and crucial for sustainable development, SDG13 – Climate action – can be introduced here at a more detailed level (Table 9.1) based on its high relevancy for the following discussion focusing on climate change policies and tourism in southern Africa.

The SDG13 target topics have been applied to tourism. For example, Target 13.1 (Table 9.1), strengthening the resilience and adaptive capacity of the sector and tourism-dependent communities to climate-related hazards and natural disasters, has been widely studied in southern Africa (see Dube et al., 2020, 2021a; Fitchett, 2021; Fitchett et al., 2016c; Mahlangu & Fitchett, 2019) as discussed in the earlier chapters. Similarly, issues that relate to understanding, knowledge, and improvement of awareness, education, and human and institutional capacity on mitigation, adaptation, and impact reduction (Target 13.3) have been directly and indirectly analysed in regional studies, especially considering tourism businesses and local communities (see Dube et al., 2020; Giddy et al., 2017; Hambira et al., 2013, 2021; Hoogendoorn et al., 2021a; Mushawemhuka et al., 2018; Pandy & Rogerson, 2018; Saarinen et al., 2012, 2020; Tervo-Kankare et al., 2018b). In addition, there have been studies on tourists' awareness, responses, and adaptation mechanisms (see Fitchett & Hoogendoorn, 2018; Friedrich et al., 2020b; Gössling et al., 2006). By contrast to the previous targets, the integration of climate change measures into national policies, strategies, and planning (Target 13.2) has been studied on a limited scale (Hambira, 2017; Hambira et al., 2020; Pandy & Rogerson, 2021), issues that will be discussed next. Furthermore, many of the indicators (Table 9.1) related to the targets of SDG13 would require major adjustments in tourism development contexts.

Climate change policies and tourism in southern Africa

The Paris Agreement and tourism futures

The Paris Agreement is an international treaty on climate change to address the continued rise in mean global temperatures (United Nations, 2015b). In principle, the treaty involves mitigation, adaptation, and finance. It was adopted in Paris in 2015 by 196 countries and came into force in 2016. The key goal of the Agreement is to limit global warming to below 2°C (preferably to 1.5°C) compared to pre-industrial levels. This would substantially reduce the impacts of climate change by the end of the 21st century. The Agreement is the outcome of a long political process (Scott et al., 2016b). It is considered as a breakthrough in global climate change governance, representing the first binding international agreement that brings all nations into a common cause to combat climate change and its effects (United Nations, 2015b).

Although climate change mitigation is considered as a collective responsibility for all, the Paris Agreement recognises that the Global North and South countries are in an uneven position with respect to climate change impacts, adaptation, and how to limit the rise of greenhouse gas emissions in the near future. The Agreement states that countries in the Global North should undertake economy-wide absolute emission reduction, while countries of the Global South "should continue enhancing their mitigation efforts, and are encouraged to move over time towards economy-wide emission reduction or limitation targets in the light of different national circumstances" (United Nations, 2015b, p. 4). Furthermore, the Agreement states that there needs to be additional financial and knowledge support provided to countries of the Global South to facilitate implementation.

As the Paris Agreement involves the need for economy-wide emissions reductions, there will be major consequences for the global tourism sector. In this respect, the Agreement will challenge future global mobilities and the common idea of viewing tourism "as a renewable resource that [...] can be utilised indefinitely" (Hollenhorst et al., 2014, p. 306). Indeed, the 'unlimited' growth that has largely characterised the past development of the tourism sector (Gössling et al., 2021; Saarinen, 2018) would cause an undesirable future regarding the impacts on climate change (Peeters et al., 2019). Thus, based on the Agreement, there is a need "to transform tourism to the low-carbon economy and improve the understanding of climate risks necessary to strengthen the climate resilience of tourism businesses and destinations" (Scott et al., 2016b, p. 940). The sector has already pledged to reduce its emissions by 70% by 2050 (Gössling & Scott, 2018), which will change the global tourism sector, its future patterns, regional structures, and flows.

International air travel alone, which is not included in the Paris Agreement, has been estimated to account for almost 2% of global emissions (Lenzen et al., 2018). Moreover, the overall tourism sector's pre-COVID-19 growth rate emissions were estimated to have increased by 130% between 2005 and 2035 (Nicholls, 2014) based on a business-as-usual scenario (i.e. no additional climate change emissions reduction control). This explains that while the sector and many policymakers are advocating a fast return to the growth path after the COVID-19 crisis, for many scholars, the crisis represents an opportunity to rethink the nature of tourism development and challenge the volume growth tourism model advocated by the mainstream industry and some policymakers such as the UNWTO (Gössling et al., 2021; Prideaux et al., 2020; Saarinen, 2021b; Sigala, 2020). As noted by Rogerson and Baum (2020) in the African context, tourism reborn in the post-COVID era (i.e. after the extended COVID crisis and associated lockdown and travel regulations) is likely to involve different forms and geographies of tourism development compared to the pre-2020 years. Some destinations may try to position themselves as low-cost and seek to get back to a growth path fast and with little concern for climate change externalities, while others may search for alternative modes and segments of tourism or an overall economic development strategy compared to the pre-COVID-19 times (Hall et al., 2020).

There are strong interconnections between the Paris Agreement and SDGs. The UNWTO and many other international development organisations expect tourism to contribute significantly (Scott et al., 2016b). In the southern African context, this can result in a policy conflict. While the climate change agreements and policies call for limiting growth and resulting emissions to achieve carbon neutrality in the future, the poverty reduction goal (SDG1), for example, most probably requires absolute economic growth in practice. It may be "possible to achieve growth without emissions" (Robinson & Shine, 2018, p. 567) but that is a highly unlikely state of business in the southern African economies on a large scale in the foreseeable future.

The new 2021 approved tourism policy for Botswana demonstrates the above-mentioned policy conflict challenge. It states that the national tourism policy is

> intended to set the scene for excellence in growth and a vibrant, globally recognised tourism industry with growth targets of 11.4% contribution to GDP by 2030 (in line with global forecasts) from the current 7.6%; increase tourism contribution to total employment from 18 000 to 40 000 by 2030.
>
> (Ministry of Environment, Natural Resources Conservation and Tourism, 2021, p. 7)

Elsewhere, the very same policy states that there is a need to "implement concrete measures in order to mitigate climate change throughout the tourism value chain" (p. 27). It is obvious that growth is needed for socioeconomic development, employment creation, and economic diversification in the country (see Hambira & Mbaiwa, 2021; Saarinen et al., 2014), for example, but at the same time, the above growth rates by 2030 create major challenges for mitigation efforts for the country's tourism industry in the context of the Paris Agreement. Furthermore, the strong growth emphasis in the tourism sector is not in line with the national climate change policy (Climate Change Response Policy, 2017) that will be discussed later.

From a global climate justice perspective, however, the policy contradictions between growth and climate change mitigation, or tourism development-based benefits for SDGs and harmful impacts for global climate change, are partly negotiable (see Beer, 2014; Schlosberg & Collins, 2014). The term climate justice has been increasingly used by governmental and non-governmental (NGOs) actors from the Global South, with a political aim to point out that there are different economic and historical responsibilities for causing global climate change. According to Fisher (2015, p. 74), this idea "has become embedded within in the United Nations Framework Convention on Climate Change through the principle of 'common but differentiated responsibilities'". Thus, from the Global South climate change policy perspective, there should be an equitable and historically and socio-economically just division of responsibility between the Global North and South in creating carbon neutrality on a global scale in the future (Schlosberg & Collins, 2014).

While the Paris Agreement recognises that the Global North and South are in different positions and capacities with regard to the mitigation process, it also highlights the need to limit carbon emissions globally (United Nations, 2015b). In this respect, one of the key strategies for the southern African tourism sector is the development of domestic and regional tourism that has a smaller carbon footprint than international long-haul aviation tourism (see Department of Tourism, 2011; Rogerson, 2016). As estimated by several scholars, the development of a more localised form of travel is likely to be an important mode in the tourism development and recovery strategies concerning the impacts of the COVID-19 pandemic (see Higgins-Desbiolles, 2020; Rogerson & Baum, 2020). This would support climate change mitigation efforts in the regional tourism industry. However, the role of domestic and regional tourism in southern Africa has been widely noted as an undervalued development strategy (see Melubo, 2020; Rogerson, 2017; Rogerson & Lisa, 2005; Rogerson & Visser, 2020). Furthermore, with some exceptions, the development of public transportation has been very limited. This includes the railway networks in the region that have rather been planned for the needs of natural resource extraction purposes than public transportation and travelling (see Marson et al., 2021).

Thus, if the growth of domestic and regional tourism is aimed to be supported for the sake of their potential benefits for the SDGs (see Ramutsindela & Mickler, 2020) and combatting climate change in southern Africa, there needs to be new mechanisms and strategies across different economic and policy sectors, including transportation, infrastructure, and border security, for example. In this respect, the role of the Southern African Development Community (SADC) is central, as it aims at achieving development, security, and economic growth to alleviate poverty and enhance the standard and quality of life of the peoples of the region (SADC, 2021). At the same time, however, there is a wider need to transform the regional tourism development in line with the global-scale zero-carbon imperative in the future, which already calls for climate change-informed policies and development strategies for the contemporary tourism sector.

National climate change policies and the role of tourism

Climate change is a significant and growing topic of concern for the southern African tourism economy and policymaking (Fitchett, 2021; Hoogendoorn

& Fitchett, 2018; Pandy & Rogerson, 2021). However, there are no explicit and detailed climate change policies for the tourism sectors across the region. Those kinds of policies are critical as they can serve as essential tools for "capacity building for tourism businesses, developers and policymakers around climate change", ensuring a deeper understanding of how the region and countries "can manage the tourism assets within their areas, and to maximise the opportunities of tourism for local economic development" (Rogerson, 2016, p. 322). Although tourism-specific climate change policies are scarce, most southern African countries have national climate change policies and strategies, which widely share an emphasis on adaptation, resilience, and capacity-building needs for future development.

South Africa has several climate change-related policy initiatives and strategies, such as the National Climate Change Response White Paper, the (draft) Climate Change Bill (Department of Environmental Affairs, 2011, 2018), and the Carbon Tax Act supporting a reduction of greenhouse emissions by using a user-pays principle (Pandy & Rogerson, 2021). The latest policy is the National Climate Change Adaptation Strategy (NCCAS) (Republic of South Africa, 2019) that was approved in 2020. It is based on the idea that climate change has the potential to reverse the gains made on the United Nations Millennium Development Goals (MDGs) and impede South Africa's ability to achieve their current SDGs. Thus, climate change is seen as representing serious risks to opportunities for socio-economic development in the country, which calls for firm adaptation strategies. The specific strategic objectives of the NCCAS are:

Objective 1: Build climate resilience and adaptive capacity to respond to climate change risk and vulnerability.

Objective 2: Promote the integration of climate change adaptation response into development objectives, policy, planning, and implementation.

Objective 3: Improve understanding of climate change impacts and capacity to respond to these impacts.

Objective 4: Ensure resources and systems are in place to enable the implementation of climate change responses.

These key objectives are categorised into several planned interventions and desired outcomes. According to Pandy and Rogerson (2021), this adaptation-focused policy places a significant emphasis on risk reduction, capacity building, and resilience. In Botswana, the Climate Change Policy (2021) also places importance on resilience and capacity building. Botswana's policy aims at mainstreaming sustainability and climate change into general development planning in the country (Ministry of Environment..., 2021a, 2021b) (Box 9.2). As a result of this, the Policy is grounded on the official National Development Plan (NDP). Furthermore, the policy promotes low-carbon development pathways and approaches that would significantly contribute to socio-economic development, environmental protection, poverty eradication, and reduction of greenhouse gases from the atmosphere. The policy also involves some sector- and theme-specific response and adaptation measures including agriculture and food security, forest and water management, and gender issues. While tourism is not highlighted as an important sector of its own in the climate change policy, the Botswana's new tourism policy necessitates the need for national and regional assessment of the effects of climate change and the potential for adaptation (Ministry of Environment..., 2021b).

BOX 9.2 POLICY RESPONSES TO CLIMATE CHANGE IN BOTSWANA

Wame L. Hambira

Botswana is endowed with natural capital that is pivotal to tourism while at the same time highly vulnerable to climate change. As a result, tourist attractions associated with vegetation, wildlife, and protected areas have been affected by climate change. Botswana's NBT has become vulnerable to climate change due to frequent droughts, erratic rainfall, shifting seasons, and extreme events such as floods and heat-waves. Droughts, for example, affect the availability of vegetation and cause wildlife to migrate in search of water which sometimes results in human-wildlife conflict.

Policy responses to climate change are therefore imperative for the economic sustainability of the tourism industry (Hambira, 2017). Policy responses comprise both mitigation and adaptation measures. The latter is more urgent for African countries such as Botswana, where climate change impacts are expected to be more severe than in any other continent. The range of adaptation measures that can be implemented are influenced by the type of tourism subsector and experiences and services offered. The measures may be technical, managerial, policy, or educational. Mitigation measures, on the other hand, are aimed at curtailing emission of greenhouse gases by the industry such as through transport or energy utilisation (Hambira et al., 2020).

SDG13 calls for urgent action to combat climate change and its impacts. Target 13b of the goal specifically calls for the integration of climate change measures into national policies, strategies, and plans. The response to climate change in Botswana is enshrined in various legislation instruments such as the National Vision 2036, the NDP 11, the Botswana National Water Policy of 2012, and the newly enacted Climate Change Policy of 2021. The climate policy actions in the sector are aimed at addressing climatic conditions such as increased temperatures, frequent droughts, and extreme climatic events such as flooding and heat-waves, and general poor climatic conditions that are unfavourable to tourist activities and experiences. The responses therefore comprise both adaptation and mitigation. Adaptation strategies include alteration of activity times in the case of high temperatures; provision of water to wildlife through manmade water holes during drought years; moving guests to alternative rooms or locations during floods; and raising awareness among both staff and guests in accommodation facilities.

With regard to mitigation, the tourism sector in Botswana has adopted predominantly technological mitigation measures and policies with the aim of lowering utility costs and improving the performance of low-carbon sources, for example, LED lighting, solar panels, room keys to operate lights, light sensors, as well as promoting energy consciousness and energy saving behaviours of both tourists and tourism sector employees. Strategies for water efficiency such as water recycling, promoting showers over bathing, and on-site water purification also count as mitigation measures. The adoption of green measures such as energy efficient fixtures and appliances, smart technologies, carbon footprint monitoring, education or awareness, and knowledge creation strategies help reduce the carbon footprint of the sector.

Even though the tourism industry in Botswana has made considerable strides in responding to climate change, the responses are often reactionary and not intentionally planned for (Saarinen et al., 2020). This may be because the adoption of sustainability practices is largely voluntary and more of a brand issue than conservation per se. It is hoped that with more educational campaigns on the tourism climate change nexus for all stakeholders, operators, and policymakers alike, responses would be more proactive.

In Namibia, the national policy is based on the general vulnerability of the country to climate change (Government of the Republic of Namibia, 2011). The overall goal of the policy is to contribute to the attainment of sustainable development in line with Namibia's Vision 2030. This goal is based on capacity and resilience-building aiming at reducing climate change risks and related shocks for people and fragile environments. Correspondingly, the Zimbabwean climate change policies involve strong emphases on capacity and awareness building, resilience, and adaptation needs (Republic of Zimbabwe, 2017).

The role of tourism is relatively invisible in the regional climate change policy landscape. In Botswana, the tourism industry is briefly mentioned in the context of agriculture and food security (agrotourism) and biodiversity conservation needs but, in general, the national climate change policies in southern Africa do not specifically include tourism and its adaptation, resilience, and capacity-building needs. In the South African context, Pandy and Rogerson (2021, p. 453) have indicated that although there is a need "for policy and action in relation to tourism and climate change", specific tourism and climate change policy development have not progressed significantly since the appearance of the initial policy discussions in the early 2010s (see Department of Tourism, 2011). According to Pandy and Rogerson (2021), this may be explained by the government's focus on other suppressing societal and political challenges and conflicts that have emerged related to transformation, equity, and social inclusivity in development (see Bhorat et al., 2020), for example. While specific political circumstances and national realities do vary in the region, the common aspect is the need to create socio-economic development and well-being for the people. Therefore, it is crucial to integrate the tourism sector and its climate change policies into wider socio-economic development processes, including SDGs initiatives, in the region. This would also position tourism-dependent communities and civil society, in general, as key stakeholders in tourism and regional development, planning, and governance thinking.

Conclusions

Resilience and capacity building towards the impacts of climate change call for good governance, innovations, and tailored policies for tourism development. Good destination governance with active innovation practices strengthens resilience and supports the needed adaptation actions and capacity building in southern Africa and together with mitigation strategies can lead tourism development towards sustainability. The challenge, however, is the implementation and how to mainstream climate change issues into tourism policies and vice versa. In this respect, integrating tourism development into national climate change policies and different sectorial climate change policies is very important. This interconnectedness would support the idea of good governance in sustainable tourism development. As Jamal and Carmargo (2018, p. 206) have noted, the definitions of tourism governance may vary but "how the state coordinates socio-economic systems to achieve sustainable tourism and how it relates to other policy actors in the process is critically important to understand". This understanding calls for further research on the nexus of climate change, tourism, and required policies in southern Africa.

Proper governance structures support destination resilience, which is a tool by which destinations in southern African could adapt to the impacts of climate change in the future. Furthermore, better integration of tourism with wider societal development policies is needed as there are great expectations for how the tourism industry could contribute to socio-economic development, employment creation, poverty reduction, and other SDGs in the region. At the same time, however, southern Africa and its regional tourism products are highly vulnerable to the estimated impacts of climate change. In this respect, a poorly climate-governed southern African tourism sector would not only remain highly vulnerable but also seriously incapacitated to deliver the expected developmental aspects to people and communities in the future.

10 Conclusions and a way forward

Following the seminal work of Preston-Whyte and Watson (2005) outlining the key threats of climate change to tourism in southern Africa, a relatively quiet decade preceded the current explosion of literature on the topic (Hoogendoorn & Fitchett, 2020). Over recent years, a significant body of work has emerged, documenting tourist and tourism operators' perceptions of climate change threats to the sector (Dube & Nhamo, 2020c; Friedrich et al., 2020a, 2020b; Hambira & Saarinen, 2015; Pandy & Rogerson, 2019), quantifying the climatic suitability of the region for tourism and trends thereof (see Fitchett et al., 2017; Mushawemhuka et al., 2020; Noome & Fitchett, 2019), empirically exploring the impacts of climate threats to tourism operations (see Dube & Nhamo, 2019a; Dube et al., 2020; Fitchett et al., 2016c; Saarinen et al., 2012), and interrogating the preparedness and adaptation strategies being adopted (see Giddy et al., 2017a; Hambira et al., 2013; Tervo-Kankare et al., 2018). There has been a less engagement on the role of tourism in contributing to climate change and on the greening of tourism operations to mitigate these impacts (see Dube & Nhamo, 2019b, 2021; Hoogendoorn et al., 2015; Machete et al., 2015).

Collectively, this growing body of research into the southern African tourism climate change nexus has revealed that while the key threats of climate change to tourism, and *vice versa*, are consistent with those reported globally, the regional setting is critical. As developing countries, scarce capital is predominantly prioritised towards infrastructural work and poverty reduction actions that meet the immediate needs of the population, rather than in preparation for the effects of climate change in decades to come (Hoogendoorn et al., 2016). The tourism sectors, which comprise an important component of the GDP of most southern African countries, are, however, already experiencing the impacts of climate-related hazards. These complexities heighten the need for on-going research into climate change and tourism in the region, critically reflecting on the balance in anticipating and planning for a changing climate. The recent publication of the 6th Assessment Report of the IPCC (2021) elevates the urgency for this sectoral-focused research.

Emerging comparatively late as a research foci, southern African tourism and climate change research has benefited from an extensive pre-developed

DOI: 10.4324/9781003102618-10

toolbox of methodologies. This has allowed for direct comparison of the findings from southern Africa with those elsewhere and for contextualisation of the severity of the threats of climate change to tourism in the region (e.g. Friedrich et al., 2020b). However, challenges relating to data availability (Fitchett et al., 2016b; Noome & Fitchett, 2019), transport to and from study sites for interviews (Mushawemhuka, 2021), and low response rates for questionnaires (Giddy et al., 2017a) have required adaptations to the research process, and for work-arounds to be developed. New methods have also emerged from southern African tourism and climate change research, such as mining TripAdvisor reviews for tourists' commentary on the weather experienced during their vacation (Fitchett et al., 2020; Fitchett & Hoogendoorn, 2019). A key consideration where adaptions are made, and new methods developed, is in weighing up the capacity for direct comparison of results between studies and applying the most suitable method to a particular tourism climate setting.

From the emerging literature on tourism and climate change, and more broadly on tourism and on climate change in southern Africa, the key climatic threats to tourism have been identified. The southern African region is projected to experience warming in the region of 1.5 times the global average (Archer et al., 2018). Both gradual climate change, manifested through increases in mean temperatures, total and seasonal rainfall, wind, humidity and storms, and extreme climate events pose threats to southern African tourism. The former will gradually affect the climatic suitability of the region for tourism, particularly as key thresholds for tourism comfort are exceeded (Fitchett et al., 2017). The latter have already started to affect the tourism sector of the region, with storms such as Tropical Storm Dando (Fitchett et al., 2016c), Tropical Cyclone Eline (Mushawemhuka, 2021), and the Cape Town 'Day Zero' drought (Dube et al., 2020) already having induced costly effects to tourism accommodation establishments and operators at tourist activities. Climate change will also have secondary impacts on tourism through changes to the distribution and phenology of fauna and flora (Mahlangu & Fitchett, 2019; Smith & Fitchett, 2020) and changes in water levels and windspeeds needed for particular activities such as white-water rafting (Giddy et al., 2017a).

These challenges require resilience-building and urgent and effective adaptation. Resilience has become one of the main ideas and approaches in tourism development thinking and how destinations can deal with change. Active resilience-building helps the tourism sector and communities in tourist destinations to manage their vulnerabilities and cope with the impacts of climate change. In this process, an adaptive capacity and related strategies are in a key role (Adger, 2009; Saarinen & Gill, 2019). Adaptation strategies vary considerably from very low-cost and often short-lived approaches such as the acquisition of fans and heaters to moderate temperatures within the rooms of tourism accommodation, through to large, costly and much longer-enduring solutions such as building storm-proof bridges, retaining walls, and pumping sea water from inundated coastal locations (Eekhout

& de Vente, 2019). Effective adaptation requires action and collaboration from individual operators through to local and regional municipalities. One approach in achieving this is the model of adaptive co-management (Armitage et al., 2008). Any adaptation requires the understanding and acknowledgement from stakeholders of the significant and urgent threat of climate change (Hoogendoorn et al., 2016). To date, adaptation strategies implemented by tourism operators have addressed climatic stressors as they arise (Hoogendoorn et al., 2016; Mushawemhuka, 2021), and municipal plans have been faced with considerable economic and political barriers (Ziervogel & Parnell, 2014).

Some of the adaptation strategies being implemented by the tourism sector serve a dual purpose in mitigating the sector's contributions to GHG emissions and alleviating the stress on water resources during drought periods. Particularly valuable have been the move to low-flow shower heads, reducing electricity and water consumption, bringing down utility costs, and minimising the added strain that the tourism sector places on climate-stressed regions (Dube et al., 2020; Hoogendoorn et al., 2015). However, as the tourism sector continues to grow, it is important for far more extensive mitigation strategies to be implemented to ensure that the tourism sector does not become responsible for its own climate change-induced demise. A range of strategies are being considered internationally, including voluntary carbon offsetting, carbon taxes, and social marketing. Locally, SATSA have recently embarked on a programme planting spekboom to sequester carbon associated with inbound flights, and following the 'Day Zero' drought (SATSA, 2019), the impacts of water restrictions coupled with extensive media communication are evidenced (Dube et al., 2020).

To develop adaptive capacity and regional and national adaptation strategies with sufficient knowledge and resourcing, sector-specific policies and governance frameworks for climate change are crucial. Recently, there have been increasing calls for climate change governance and policy frameworks for sustainable and climate-concerned tourism development in southern Africa (see Hambira et al., 2020; Hambira & Saarinen, 2015; Pandy & Rogerson, 2021). However, while there are many national climate change policies and action plans in the region, the tourism sector-specific policies and strategies are largely absent. These policies are required as they are tools for resilience and adaptive capacity building for the tourism sector, local and regional developers, and tourism-dependent communities (Rogerson, 2016; Saarinen et al., 2020).

Policymaking and governance calls in southern Africa are often linked with a need to develop good governance. In tourism, this good governance refers to frameworks advocated for the sector to create economic and social benefits beyond the industry's own operations (Siakwah et al., 2020). This integrates good governance for sustainable development thinking in tourism and questions such as how the tourism industry could contribute to SDGs in the region. To deliver development and well-being, the tourism sector-specific climate change policies and governance frameworks need to be firmly integrated into wider societal development strategies, politics, and processes in southern Africa. Otherwise, we may end up developing and governing climate change resilient tourism for the sake of the industry alone, without creating those expected and actively promoted socio-economic benefits for surrounding societies and environments.

The relationships between climate change and tourism, and questions on how tourism could be more resilient, adaptive, and better embedded with wider societal development needs, are highly complex and fluid. To understand these relations and issues, however, we need to emphasise the context in which tourism operates and is aimed to be developed and used for realising various goals, including sustainability. Thus, there is a real need for further research on tourism and climate change, specifically in southern Africa. At a general level, more research is needed on the impacts of climate change on different forms of tourism in the region, including places and countries that have not yet been the focus of academic research. Moreover, tourism and climate change research should also focus on the ecological and social environments where tourism takes place and their specific characteristics supporting different kinds of tourism activities. This would help in reducing the vulnerability of tourism and develop better-informed adaptation strategies that are tailored to the needs of specific tourism destinations.

Future research avenues should also consider the extent to which different governance frameworks and tourism policies adequately anticipate climate change impacts on tourism in the region. As noted by Adu-Ampong et al. (2021) in the Sub-Saharan African context, this line of research should also focus on how tourism policies provide support for appropriate adaptation strategies for destinations with different levels of vulnerability, resources, adaptive capacity, knowledge, and resilience. There is also a need to create better and wider awareness of the impacts and process of climate change and what potential effects they will have on tourism and surrounding socio-ecological systems. Furthermore, it is important to create awareness in the tourism sector on how to support mitigation efforts and other commitments based on the Paris Agreement. This calls for increasing education and training on climate change and tourism relations. In this work, the university programmes and scholars play a pivotal role together with other higher education institutions and actors. Overall, the way forward with tourism and climate change relations in southern Africa is based on the creation of active research programmes in different countries, development of climate change-informed tourism education and policies, and effective dissemination of knowledge. This book aims to serve all these core needs by analysing and creating the ground knowledge of the past tourism and climate change research in southern Africa.

References

Abba Omar, S., & Abiodun, B.J. (2020). Characteristics of cut-off lows during the 2015–2017 drought in the Western Cape, South Africa. *Atmospheric Research, 235*(1), 104772.

Abegg, B., & Elsasser, H. (1996). Climate, weather, and tourism in the Swiss Alps. *Geographische Rundschau, 48*, 737–742.

Abegg, B., König, U., Bürki, R., & Elsasser, H. (1997). Climate impact assessment and tourism. *Erde, 128*, 105–116.

Abiodun, B.J., Makhanya, N., Petja, B., Abatan, A.A., & Oguntunde, P.G. (2019). Future projection of droughts over major river basins in Southern Africa at specific global warming levels. *Theoretical and Applied Climatology, 137*(3), 1785–1799.

Abraham, J.O., Hempson, G.P., & Staver, A.C. (2019). Drought-response strategies of savanna herbivores. *Ecology and Evolution, 9*(12), 7047–7056.

Abrahams, Z., Hoogendoorn, G., & Fitchett, J.M. (2021). Glacier tourism and tourist reviews: An experiential engagement with the concept of "last chance tourism". *Scandinavian Journal of Hospitality and Tourism*. https://doi.org/10.1080/15022250.2021.1974545

Adger, W.N. (2000). Social and ecological resilience: Are they related? *Progress in Human Geography, 24*, 347–364.

Adger, W.N. (2006). Vulnerability. *Global Environmental Change, 16*(3), 268–281.

Adger, W.N. (2009). Social capital, collective action, and adaptation to climate change. *Economic Geography, 79*(4), 387–404.

Adger, W.N., Huq, S., Brown, K., Conway, D., & Hulme, M. (2003). Adaptation to climate change in the developing world. *Progress in Developing Studies, 3*(3), 179–195.

Adhikari, M. (2010). A total extinction confidently hoped for: The destruction of Cape San society under Dutch colonial rule, 1700–1795. *Journal of Genocide Research, 12*(1–2), 19–44.

Adu-Ampong, E.A., Kimbu, A.N., & Saarinen, J. (2021). Reconsidering sustainability in tourism policy and planning in Sub-Saharan Africa: A research agenda. In E.A. Adu-Ampong, & A.N. Kimbu (Eds.), *Sustainability in tourism policy and planning in Sub-Saharan Africa: The past, the present and the future* (pp. 175–186). Routledge.

African Development Bank Group. (2021, August 17). *Mauritius Economic Outlook.* https://www.afdb.org/en/countries/southern-africa/mauritius/mauritius-economic-outlook

Afriski Mountain Resort. (2019). *Home – Afriski Mountain Resort*. https://www. afriski.net/

Agüero-Torales, M.M., Cobo, M.J., Herrera-Viedma, E., & López-Herrera, A.G. (2019). A cloud-based tool for sentiment analysis in reviews about restaurants on TripAdvisor. *Procedia Computer Science, 162*, 392–399.

Ahmed, F., Sookrajh, R., & Moodley, V. (2008). The environmental impacts of beach sport tourism events: A case study of the Mr. Price Pro surfing event, Durban, South Africa. *Africa Insight, 38*(3), 73–85.

Akadiri, S.S., Lasisi, T.T., Uzuner, G., & Akadiri, A.C. (2019b). Examining the impact of globalization in the environmental Kuznet curve hypothesis: The case of tourist destination states. *Environmental Science and Pollution Research, 26*, 12605–12615.

Akadiri, S.S., Lasisi, T.T., Uzuner, G., & Akadiri, A.C. (2020). Examining the causal impacts of tourism, globalization, economic growth and carbon emissions in tourism island territories: Bootstrap panel Granger causality analysis. *Current Issues in Tourism, 23*(4), 470–484.

Alim, M.A., Rahman, A., Tao, Z., Samali, B., Khan, M.M., & Shirin, S. (2020). Feasibility analysis of a small-scale rainwater harvesting system for drinking water production in Werrington, New South Wales, Australia. *Journal of Cleaner Production, 270*(10), 122437.

Allen, R.B. (2001). Licentious and unbridled proceedings: The illegal slave trade to Mauritius and the Seychelles during the early nineteenth century. *The Journal of African History, 42*(1), 91–116.

Alonso, A.D., & Ogle, A. (2008). Exploring design among small hospitality and tourism operations. *Journal of Retail and Leisure Property, 7*(4), 325–337.

Al-Saqaf, W., & Berglez, P. (2019). How do social media users link different types of extreme events to climate change? A study of Twitter during 2008–2017. *Journal of Extreme Events, 6*(2), 1950002

Altman, J., Ukhvatkina, O.N., Omelko, A.M., Macek, M., Plener, T., Pejcha, V., Cerny, T., Petrik, P., Srutek, M., Song, J.S., & Zhmerenetsky, A.A. (2018). Poleward migration of the destructive effects of tropical cyclones during the 20th century. *Proceedings of the National Academy of Sciences, 115*(45), 11543–11548.

Amaral, F., Tiago, T., & Tiago, F. (2014). User-generated content: Tourists' profiles on TripAdvisor. *International Journal on Strategic Innovative Marketing, 1*, 137–147.

Amusan, L., & Olutola, O. (2017). Climate change and sustainable tourism: South Africa caught in-between. *African Journal of Hospitality, Tourism and Leisure, 6*(4), 1–15.

Anisimov, A., Magnan, A.K., & Duvat, V.K.E. (2020). Learning from risk reduction pilot projects for enhancing long-term adaptation governance: The case of Mauritius Island (Indian Ocean). *Environmental Science & Policy, 108*, 93–103.

Archer, E.R.M. (2019). Learning from South Africa's recent summer rainfall droughts: How might we think differently about response? *Area, 51*(3), 603–608.

Archer, E.R.M., Engelbrecht, F.A., Hänsler, A., Landman, W., Tadross, M., & Helmschrot, J. (2018). Seasonal prediction and regional climate projections for southern Africa. In R. Revermann, K.M. Krewenka, U. Schmiedel, J.M. Olwoch, J. Helmschrot, & N. Jürgens (Eds.), *Climate change and adaptive land management in southern Africa* (pp. 14–21), Klaus Hess Publishers.

Archibald, S. (2016). Managing the human component of fire regimes: Lessons from Africa. *Philosophical Transactions B, 371*, 20150346.

Armitage, D., Marschke, M., & Plummer, R. (2008). Adaptive co-management and the paradox of learning. *Global Environmental Change, 18*, 86–98.

ASSAR (Adaptation at Scale in Semi-Arid Regions). (2021, April 17). *What global warming of 1.5°C and higher means for Namibia (2018)*. http://www.assar.uct. ac.za/sites/default/files/image_tool/images/138/1point5degrees/ASSAR_Namibia_global_warming.pdf.

Azam, M., Alam, M., & Hafeez, M.H. (2018). Effect of tourism on environmental pollution: Further evidence from Malaysia, Singapore and Thailand. *Journal of Cleaner Production, 190*, 330–338.

Baffi, S., Donaldson, R., & Spocter, M. (2020). Tourist mobilities in Cape Town: Unveiling practices in the post-apartheid city. *Tourism Planning and Development, 17*(5), 537–555.

Baggio, R., Scott, N., & Cooper, C. (2010). Improving tourism destination governance: A complexity science approach. *Tourism Review, 65*(4), 51–60.

Baird, J., Plummer, R., & Bodin, Ö. (2016). Collaborative governance for climate change adaptation in Canada: Experimenting with adaptive co-management. *Regional Environmental Change, 16*(3), 747–758.

Baker, M., & Mearns, K.F. (2017). Applying sustainable tourism indicators to measure the sustainability performance of two tourism lodges in the Namib Desert. *African Journal of Hospitality, Tourism and Leisure, 6*(2), 1–22.

Balsalobre-Lorente, D., Driha, O.M., Shahbaz, M., & Sinha, A. (2020). The effects of tourism and globalization over environmental degradation in developed countries. *Environmental Science and Pollution Research, 27*, 7130–7144.

Banerjee, S., & Chua, A.Y. (2016). In search of patterns among travellers' hotel ratings in TripAdvisor. *Tourism Management, 53*, 125–131.

Barimalala, R., Blamey, R.C., Desbiolles, F., & Reason, C.J.C. (2020). Variability in the Mozambique Channel trough and impacts on southeast African rainfall. *Journal of Climate, 33*(2), 749–765.

Barimalala, R., Desbiolles, F., Blamet, R.C., & Reason, C. (2018). Madagascar influence on the south Indian Ocean Convergence Zone, the Mozambique channel trough and southern African Rainfall. *Geophysical Research Letters, 45*, 11380–11389.

Barnard, A. (2019). *Bushmen: Kalahari hunter-gatherers and their descendants*. Cambridge University Press.

Baudoin, M., Vogel, C., Nortje, K., & Nai, M. (2017). Living with drought in South Africa: Lessons learnt from the recent El Niño drought period. *International Journal Disaster Risk Reduction, 23*, 28–137.

Baumber, A., Merson, J., & Lockhart Smith, C. (2021). Promoting low-carbon tourism through adaptive regional certification. *Climate, 9*(15). https://doi.org/10.3390/cli9010015

Beck, C., Grieser, J., Kottek, M., Rubel, F., & Rudolf, B. (2005). Characterizing global climate change by means of Köppen Climate Classification. *Klimatstatusbericht, 10*, 139–149.

Beck, H.E., Zimmermann, N.E., McVicar, T.R., Vergopolan, N., Berg, A., & Wood, E.F. (2018). Present and future Köppen-Geiger climate classification maps at 1-km resolution. *Scientific Data, 5*, 180214.

Becken, S. (2002). Analysing international tourist flows to estimate energy use associated with air travel. *Journal of Sustainable Tourism, 10*(2), 114–131.

Becken, S. (2005). Harmonising climate change adaptation and mitigation: The case of tourist resorts in Fiji. *Global Environmental Change, 15,* 381–393.

Becken, S. (2010). *The importance of climate and weather for tourism.* Land Environment & People.

Becken, S. (2013). A review of tourism and climate change as an evolving knowledge domain. *Tourism Management Perspectives, 6,* 53–62.

Becken, S. (2017). Evidence of a low-carbon tourism paradigm? *Journal of Sustainable Tourism, 25,* 832–850.

Becken, S., & Hay, J. (2007). *Tourism and climate change: Risk and opportunities.* Channelview Publications.

Becken, S., Whittlesea, E., Loehr, J., & Scott, D. (2020). Tourism and climate change: Evaluating the extent of policy integration. *Journal of Sustainable Tourism, 28*(10), 1603–1624.

Becken, S., & Wilson, J. (2013). The impacts of weather on tourist travel. *Tourism Geographies, 15*(4), 620–639.

Becker, S. (1998). Beach comfort index: A new approach to evaluate the thermal conditions of beach holiday resort using a South Africa example. *GeoJournal, 44*(4), 297–307.

Beer, C.T. (2014). Climate justice, the Global South, and policy preferences of Kenyan environmental NGOs. *The Global South, 8*(2), 84–100.

Belén Gómez Martín, M. (2005). Weather, climate and tourism a geographical perspective. *Annals of Tourism Research, 32*(3), 571–591.

Berkes, F., & Ross, H. (2013). Community resilience: Toward an integrated approach. *Society and Natural Resources, 26,* 5–20.

Betzold, C., & Mohamed, I. (2017). Seawalls as a response to coastal erosion and flooding: A case study from Grande Comore, Comoros (West Indian Ocean). *Regional Environmental Change, 17*(4), 1077–1087.

Bewell, A. (2004). Romanticism and colonial natural history. *Studies in Romanticism, 43*(1), 5–34.

Bhorat, H., Lilenstein, K., Oosthuizen, M., & Thornton, A. (2020). Structural transformation, inequality, and inclusive growth in South Africa. *WIDER Working Paper,* 2020/50.

Bickford-Smith, V. (2009). Creating a city of the tourist imagination: The case of Cape Town, 'The fairest cape of them all'. *Urban Studies, 46*(9), 1763–1785.

Bigano, A., Hamilton, J.M., & Tol, R.S.J. (2006). The impact of climate on holiday destination choice. *Climatic Change, 76*(3–4), 389–406.

Biggs, D., Hall, C.M., & Stoeckl, N. (2012). The resilience of formal and informal tourism enterprises to disasters: Reef tourism in Phuket, Thailand. *Journal of Sustainable Tourism, 20*(5), 645–665.

Booyens, I., & Visser, G. (2010). Tourism SMME development on the urban fringe: The case of Parys, South Africa. *Urban Forum, 21,* 367–385.

Bossio, C.M., Ford, J., & Labbé, D. (2019). Adaptive capacity in urban areas of developing countries. *Climatic Change, 157,* 279–297.

Bramwell, B. (2011). Governance, the state and sustainable tourism: A political economy approach. *Journal of Sustainable Tourism, 19*(4–5), 459–477.

Bramwell, B., Higham, J., Lane, B., & Miller, G. (2017). Twenty-five years of sustainable tourism and the Journal of Sustainable Tourism: Looking back and moving forward. *Journal of Sustainable Tourism, 25*(1), 1–9.

Bramwell, B., & Lane, B. (2011). Critical research on the governance of tourism and sustainability. *Journal of Sustainable Tourism, 19*(4–5), 411–421.

Brick, K., & Visser, M. (2018). *Green nudges in the demand-side management (DSM) toolkit: Evidence from drought-stricken Cape Town*. Draft Working Paper, https://doi.org/10.13140/RG.2.2.16413.00489

Broadbent, A. (2013). *Philosophy of epidemiology*. Palgrave Macmillan.

Brooks, N., Clarke, J., Ngaruiya, W.G., & Wangui, E.E. (2020). African heritage in a changing climate. *Azania: Archaeological Research in Africa, 55*(3), 297–328.

Brown, K. (2011). Sustainable adaptation: An oxymoron? *Climate and Development, 3*(1), 21–31.

Buckley, R. (2017). Perceived resource quality as a framework to analyze the impacts of climate change adventure tourism: Snow, surf, wind and whitewater. *Tourism Review International, 21*(3), 241–254.

Burls, N.J., Blamey, R.C., Cash, B.A., Swenson, E.T., al Fahad, A., Bopape, M-J.M., Straus, D.M., & Reason, C.J.C. (2019). The Cape Town "Day Zero" drought and Hadley cell expansion. *NPJ Climate and Atmospheric Science, 2*, 27.

Buso, T., Dell'Anna, F., Becchio, C., Bottero, M.C., & Corgnati, S.P. (2017). Of comfort and cost: Examining indoor comfort conditions and guests' valuations in Italian hotel rooms. *Energy Research and Social Science, 32*, 94–111.

Butler, R.W. (1999). Sustainable tourism: A state-of-the-art review. *Tourism Geographies, 1*, 7–25.

Butler, R.W. (2017). *Tourism and resilience*. CABI.

Byakatonda, J., Parida, B.P., Kenabatho, P.K., & Maolafhi, D.B. (2018a). Analysis of rainfall and temperature time series to detect long-term climatic trends and variability over semi-arid Botswana. *Journal of Earth System Science, 127*(25), 1–20.

Byakatonda, J., Parida, B.P., Moalafhi, D.B., & Kenabatho, P.K. (2018b). Analysis of long term drought severity characteristics and trends across semiarid Botswana using two drought indices. *Atmospheric Research, 213*, 492–508.

Calgaro, E., Lloyd, K., & Dominey-Howes, D. (2014). From vulnerability to transformation: A framework for assessing the vulnerability and resilience of tourism destinations. *Journal of Sustainable Tourism, 22*(3), 341–360.

Campos Rodrigues, L., Freire-González, J., González Puig, A., & Puig-Ventosa, I. (2018). Climate change adaptation of alpine ski tourism in Spain. *Climate, 6*(2), 29.

Cape Town Tourism. (2018). Data and insights, April 2018. https://www.capetown.travel/wp-content/uploads/2018/06/CTT-Research-Report-April-2018_Classic.pdf

Carmody, P. (2019). *Development theory and practice in a changing world*. Routledge.Carolli, M., Zolezzi, G., Geneletti, D., Siviglia, A., Carolli, F., & Cainelli, O. (2017). Modelling white-water rafting suitability in a hydropower regulated Alpine River. *Science of the Total Environment, 579*(1), 1035–1049.

Cathala, C., Núñez, A., & Rios, A.R. (2018). Water in the time of drought: Lessons from five droughts around the World (Policy Brief IDB-PB-295). Inter-American Development Bank. https://webimages.iadb.org/publications/2018-12/Water-in-the-time-of-drought-Lessons-from-five-droughts-around-the-world.pdf?VqtTtKrv8tcdz0Aeuu_4h3KFVb6piXFg

Cecinati, F., Matthews, T., Natarajan, S., McCullen, N., & Coley, D. (2019). Mining social media to identify heat waves. *International Journal of Environmental Research and Public Health, 16*(5), 762.

Chen, J., Zhao, A., Zhao, Q., Song, M., Baležentis, T., & Streimikiene, D. (2018). Estimation and factor decomposition of carbon emissions in China's Tourism sector. *Problems of Sustainable Development, 13*(2), 91–101.

Chen, L.-F. (2019). Green certification, e-commerce, and low-carbon economy for international tourist hotels. *Environmental Science and Pollution Research, 26*, 17965–17973.

Chigora, F., Nyoni, T., & Mutambara, E. (2019). Forecasting CO2 wemission for Zimbabwe's tourism destination vibrancy: A univariate approach using Box-Jenkins ARIMA model. *African Journal of Hospitality, Tourism and Leisure, 8*(2), 1–15.

Chishti, M.Z., Ullah, S., Ozturk, I., & Usman, I. (2018). Examining the asymmetric effects of globalization and tourism on pollution emissions in South Asia. *Environmental Science and Pollution Research, 27*, 27721–27737.

Chiutsi, S., & Saarinen, J. (2017). Local participation in transfrontier tourism: The case of Sengwe community in Great Limpopo Transfrontier Conservation Area, Zimbabwe. *Development Southern Africa, 34*(3), 260–275.

Christie, I., Fernandes, E., Messerli, H., & Twining-Ward, L. (2013). *Tourism in Africa. Harnessing tourism for growth and improved livelihoods*. The World Bank.

Christopher, A.J. (1988). Roots of urban segregation: South African at Union, 1910. *Journal of Historical Geography, 14*(2), 151–169.

Christopherson, S., Michie, J., & Tyler, P. (2010). Regional resilience: Theoretical and empirical perspectives. *Cambridge Journal of Regions, Economy and Society, 3*(1), 3–10.

Chung, M.G., Dietz, T., & Liu, J. (2018). Global relationships between biodiversity and nature-based tourism in protected areas. *Ecosystem Services, 34*, 11–23.

Cochrane, J. (2010). The sphere of tourism resilience. *Tourism Recreation Research, 35*(2), 173–185.

Coghlan, A., & Prideaux, B. (2009). Welcome to the wet tropics: The importance of weather in reef tourism resilience. *Current Issues in Tourism, 12*(2), 89–104.

Cole, S. (2014). Tourism and water: From stakeholders to rights holders, and what tourism businesses need to do. *Journal of Sustainable Tourism, 22*(1), 89–106.

Cole, S., & Ferguson, L. (2015). Towards a gendered political economy of water and tourism. *Tourism Geographies, 17*(4), 511–528.

Cook, C., Reason, C.J.C., & Hewitson, B.C. (2004). Wet and dry spells within particularly wet and dry summers in the South African summer rainfall region. *Climate Research, 26*, 17–31.

Cornelissen, S. (2017). *The global tourism system: Governance, development and lessons from South Africa*. Routledge.

Cowling, R.M., Esler, K.J., & Rundel, P.W. (1999). Namaqualand, South Africa – An overview of a unique winter-rainfall desert ecosystem. *Plant Ecology, 142*(1–2), 3–21.

Crétat, J., Richard, Y., Pohl, B., Rouault, M., Reason, C., & Fauchereau, N. (2012). Recurrent daily rainfall patterns over South Africa and associated dynamics during the core of the austral summer. *International Journal of Climatology, 32*, 261–273.

Crutzen, P.J. (2002). Geology of mankind. *Nature, 415*, 23.

Dantas, É.S., & Mather, C. (2011). Ilha de Moçambique: Conserving and managing world heritage in the developing world. *Tourism Review International, 15*(1–2), 51–62.

Darkoh, M.B.K., Khayesi, M., & Mbaiwa, J.E. (2014). Impacts and responses to climate change at the micro-spatial scale in Malawi, Botswana and Kenya. In M.A.M. Salih (Ed.), *Local climate change and society* (pp. 109–124). Routledge.

Davidson, D.J. (2010). The applicability of the concept of resilience to social systems: Some sources of optimism and nagging doubts. *Society and Natural Resources, 23*, 1135–1149.

Davoudi, S. (2012). Resilience: A bridging concept or a dead end? *Planning Theory & Practice, 13*(2), 299–307.

Dedering, T. (2000). The Ferreira Raid of 1906: Boers, Britons and Germans in Southern Africa in the aftermath of the South African war. *Journal of Southern African Studies, 26*(1), 43–59.

de Freitas, C.R. (1990). Recreation climate assessment. *International Journal of Climatology, 10*, 89–103.

de Freitas, C.R. (2003). Tourism climatology: Evaluating environmental information for decision making and business planning in the recreation and tourism sector. *International Journal of Biometeorology, 48*, 45–54.

de Freitas, C.R., Scott, D., & McBoyle, G. (2008). A second-generation climate index for tourism (CIT): Specification and Verification. *International Journal of Biometeorology, 52*, 399–407.

Denstadli, J.M., Jacobsen, J.K.S., & Lohmann, M. (2011). Tourist perceptions of summer weather in Scandinavia. *Annals of Tourism Research, 38*(3), 920–940.

Department of Environmental Affairs. (2011). *National climate change response white paper*. Department of Environmental Affairs.

Department of Environmental Affairs. (2018). *Draft climate change bill*. Department of Environmental Affairs.

Department of Environmental Affairs and Tourism [DEAT]. (2007). *Building the tourism economy, going for gold*. DEAT.

Department of Tourism. (2011). *Draft national tourism and climate change action plan*. Department of Tourism.Desbiolles, F., Blamey, R., Illig, S., James, R., Barimalala, R., Renault, L., & Reason, C. (2018). Upscaling impact of wind/sea surface temperature mesoscale interactions on southern African austral summer climate. *International Journal of Climatology, 38*, 4651–4660.

Desbiolles, F., Howard, E., Blamey, R.C., Barimalala, R., Hart, N.C.G., & Reason, C.J.C. (2020). Role of ocean mesoscale structures in shaping the Angola-Low pressure system and the southern African rainfall. *Climate Dynamics, 54*, 3685–3704.

Devi, S. (2019). Cyclone Idai: 1 month later, devastation persists. *The Lancet, 393*(10181), 1585.

Dillimono, H.D., & Dickinson, J.E. (2015). Travel, tourism, climate change, and behavioral change: Travellers' perspectives from a developing country, Nigeria. *Journal of Sustainable Tourism, 23*(3), 437–454.

Di Minin, E., Slotow, R., Hunter, L.T., Pouzols, F. M., Toivonen, T., Verburg, V.H., & Moilanen, A. (2016). Global priorities for national carnivore conservation under land use change. *Scientific Reports, 6*, 23814.

Dogru, T., Elizabeth, A., Umit Bulut, M., & Suess, C. (2019). Climate change: Vulnerability and resilience of tourism and the entire economy. *Tourism Management, 72*, 292–305.

Donaldson, R., & Ferreira, S. (2009). (Re-)creating urban destination image: Opinions of foreign visitors to South Africa on safety and security? *Urban Forum, 20*(1), 1–18.

Donat, M.G., Alexander, L.V., Yang, H., Durre, I., Vose, R., & Caesar, J. (2013). Global land-based datasets for monitoring climatic extremes. *Bulletin of the American Meteorological Society, 94*(7), 997–1006.

Donnelly, C., Ernst, K., & Arheimer, B. (2018). A comparison of hydrological climate services at different scales by users and scientists. *Climate Services, 11*, 24–35.

Driscoll, D.L. (2011). Introduction to primary research: Observations, surveys and interviews. In C. Lowe, & P. Zemliansky (Eds.), *Writing spaces: Readings on writings, vol. 2* (pp. 153–174). Parlor Press.

Dube, K., & Mearns, K. (2019). Tourism and recreational potential of green building a case study of Hotel Verde Cape Town. In G. Nhamo, & V. Mjimba (Eds.), *The green building evolution* (pp. 200–219). Africa Institute of South Africa.

Dube, K., Mearns, K., Mini, S., & Chapungu, L. (2018). Tourists' knowledge and perceptions on the impact of climate change on tourism in Okavango Delta, Botswana. *African Journal of Hospitality, Tourism and Leisure, 7*(4), 1–18.

Dube, K., & Nhamo, G. (2019a). Climate change and potential impacts on tourism: Evidence from the Zimbabwean side of the Victoria Falls. *Environment, Development and Sustainability, 21*, 2025–2041.

Dube, K., & Nhamo, G. (2019b). Climate change and the aviation sector: A focus on the Victoria Falls tourism route. *Environmental Development, 29*, 5–15.

Dube, K., & Nhamo, G. (2020a). Evidence and impact of climate change on South African national parks. Potential implications for tourism in the Kruger National Park. *Environmental Development, 33*, 100485.

Dube, K., & Nhamo, G. (2020b). Vulnerability of nature-based tourism to climate variability and change: Case of Kariba resort town, Zimbabwe. *Journal of Outdoor Recreation and Tourism, 29*, 100281.

Dube, K., & Nhamo, G. (2020c). Tourism business operators' perceptions, knowledge and attitudes towards climate change at Victoria Falls. *TD: The Journal for Transdisciplinary Research in Southern Africa, 16*(1), 1–10.

Dube, K., & Nhamo, G. (2020d). Tourist perceptions and attitudes regarding the impacts of climate change on Victoria Falls. *Bulletin of Geography. Socio-Economic Series, 47*, 27–44.

Dube, K., & Nhamo, G. (2020e). Sustainable development goals localization in the tourism sector: Lessons from Grootbos Private Nature Reserve, South Africa. *GeoJournal.* https://doi.org/10.1007/s10708-020-10182-8

Dube, K., Nhamo, G., & Chikodzi, D. (2020). Climate change-induced droughts and tourism: Impacts and responses of Western Cape province, South Africa. *Journal of Outdoor Recreation and Tourism.* https://doi.org/10.1016/j.jort.2020.100319

Dube, K., & Nhamo, G. (2021). Greenhouse gas emissions and sustainability in Victoria Falls: Focus on hotels, tour operators and related attractions. *African Geographical Review, 40*(2), 125–140.

Dube, K., Nhamo, G., & Chikodzi, D. (2021a). Rising sea level and its implications on coastal tourism development in Cape Town, South Africa. *Journal of Outdoor Tourism and Recreation, 33*, 1–13.

Dube, K., Nhamo, G., & Chikodzi, D. (2021b). COVID-19 pandemic and prospects for recovery of global aviation industry. *Journal of Air Transport Management, 92*, 102022.

Dube, O.P. (2003). Impact of climate change, vulnerability and adaptation options: Exploring the case of Botswana through Southern Africa: A Review. *Botswana Notes and Records, 35*, 147–168.

Dubois, G., & Ceron, J.-P. (2006). Tourism and climate change: Proposals for a research agenda. *Journal of Sustainable Tourism, 14*(4), 399–415.

Dunning, C.M., Black, E.C., & Allan, R.P. (2018). Later wet seasons with more intense rainfall over Africa under future climate change. *Journal of Climate, 31*(23), 9719–9738.

Duval, D.T. (2013). Critical issues in air transport and tourism. *Tourism Geographies, 15*(3), 494–510.

Duvat, V.K., Anisimov, A., & Magnan, A.K. (2020). Assessment of coastal risk reduction and adaptation-labelled responses in Mauritius Island (Indian Ocean). *Regional Environmental Change, 20*, 110.

Eagle, L., Hamann, M., & Low, D.R. (2016). The role of social marketing, marine turtles and sustainable tourism in reducing plastic pollution. *Marine Pollution Bulletin, 107*, 324–332.

Eckert, E. (2020). *Assessment of sustainable tourism development in Windhoek, Namibia: Development and implementation of an adapted criteria catalogue in line with respective local conditions.* Masters Dissertation Submitted to the University of Applied Sciences, Bremen.

Eekhout, J.P.C., &d de Vente, J. (2019). Assessing the effectiveness of sustainable land management for large-scale climate change adaptation. *Science of the Total Environment, 654*, 85–93.

Emerton, R., Cloke, H., Ficchi, A., Hawker, L., de Wit, S., Speight, L., Prudhomme, C., Rundell, P., West, R., Neal, J., & Cuna, J. (2020). Emergency flood bulletins for cyclones Idai and Kenneth: A critical evaluation of the use of global flood forecasts for international humanitarian preparedness and response. *International Journal of Disaster Risk Reduction, 50*, 101811.

Engelbrecht, C., & Engelbrecht, F. (2016). Shifts in Köppen-Geiger climate zones over southern Africa in relation to key global temperature goals. *Theoretical and Applied Climatology, 123*, 247–261.

Engelbrecht, C.J., Landman, W.A., Engelbrecht, F.A., & Malherbe, J. (2015a). A synoptic decomposition of rainfall over the Cape south coast of South Africa. *Climate Dynamics, 44*, 2589–2607.

Engelbrecht, F., Adegoke, J., Bopape, M-J., Naidoo, M., Garland, R., Thatcher, M., McGregor, J., Katzfey, J., Werner, M., Ichoku, C., & Gatebe, C. (2015b). Projections of rapidly rising surface temperatures over Africa under low mitigation. *Environmental Research Letters, 10*(8), 085004.

Engelbrecht, F.A., Landman, W.A., Engelbrecht, C.J., Landman, S., Bopape, M-J.M., Roux, B., McGregor, J.L., & Thatcher, M. (2011). Multi-scale climate modelling over southern Africa using a variable-resolution global model. *Water SA, 37*(5), 647–658.

Engelbrecht, F.A., McGregor, J.L., & Engelbrecht, C.J. (2009). Dynamics of the conformal-cubic atmospheric model projected climate-change signal over southern Africa. *International Journal of Climatology, 29*, 1013–1033.

Enqvist, J.P., & Ziervogel, G. (2019). Water governance and justice in Cape Town: An overview. *WIREs Water, 6*(4), 1–15.

Espiner, S., & Becken, S. (2014). Tourist town on the edge: Conceptualizing vulnerability and resilience in a protected area tourism. *Journal of Sustainable Tourism, 22*(4), 646–665.

Espiner, S., Orchiston, C., & Higham, J. (2017). Resilience and sustainability: A complementary relationship? Towards a practical conceptual model for the sustainability-resilience nexus in tourism. *Journal of Sustainable Tourism, 25*(10), 1385–1400.

Eswatini Tourism Statistics. (2019). *Eswatini Tourism Statistics -arrivals by country.* https://www.thekingdomofeswatini.com/wp-content/uploads/2019/09/8-August-Arrivals-19.pdf

eThekwini Municipality. (2021). *Integrated development plan. 5 year plan: 2017/2018–2021/2022.* eThekwini Municipality, Durban.

Ezeuduji, I.O., & Nkosi, G.S. (2017). Tourism destination competitiveness using brand essence: Incorporating the 'zuluness' of the Zulu Kingdom. *African Journal of Hospitality, Tourism and Leisure, 6*(1), 1–10.

Fairer-Wessels, F.A. (2017). Determining the impact of information on rural livelihoods and sustainable tourism development near protected areas in Kwa-Zulu Natal, South Africa. *Journal of Sustainable Tourism, 25*(1), 10–25.

Fang, Y., & Yin, J. (2015). National assessment of climate resources for tourism seasonality in China using the tourism climate index. *Atmosphere, 6*(2), 183–194.

Fang, Y., Yin, J., & Wu, B. (2018). Climate change and tourism: A scientometric analysis using Citespace. *Journal of Sustainable Tourism, 26*(1), 108–126.

Fedderke, J.W. (2018). Exploring unbalanced growth: Understanding the sectoral structure of the South African economy. *Economic Modelling, 72,* 177–189.

Ferreira, S.L.A., & Harmse, A. (2014). Kruger National Park: Tourism development and issues around the management of large of tourists. *Journal of Ecotourism, 13*(1), 16–34.

Fisher, S. (2015). The emerging geographies of climate justice. *The Geographical Journal, 181*(1), 73–82.

Fitchett, A. (2014). Adaptive co-management in the context of informal settlements. *Urban Forum, 25*(3), 355–374.

Fitchett, A. (2017). SuDS for managing surface water in Diepsloot informal settlement, Johannesburg, South Africa. *Water SA, 43*(2), 310–322.

Fitchett, A., Govender, P., & Vallabh, P. (2020a). An exploration of green roofs for indoor and exterior temperature regulation in the South African interior. *Environment, Development and Sustainability, 22,* 5025–5044.

Fitchett, J.M. (2018). Recent emergence of CAT5 tropical cyclones in the South Indian Ocean. *South African Journal of Science, 114*(11–12), 1–6.

Fitchett, J.M. (2019). The holocene climates of South Africa. In J. Knight, & C.M. Rogerson (Eds.), *The geography of South Africa: Contemporary changes and new directions* (pp. 47–56). Springer.

Fitchett, J.M. (2021). Climate change threats to urban tourism in South Africa. In C.M. Rogerson, & J.M. Rogerson (Eds.), *Urban tourism in the global south* (pp. 77–92). Springer.

Fitchett, J.M., Fortune, S.M., & Hoogendoorn, G. (2020b). Tourists' reviews of weather in five Indian Ocean islands. *Singapore Journal of Tropical Geography, 41*(2), 171–189.

Fitchett, J.M., & Grab, S.W. (2014). A 66-year tropical cyclone record for south-east Africa: Temporal trends in a global context. *International Journal of Climatology, 34*(13), 3604–3615.

Fitchett, J.M., Grab, S.W., & Portwig, H. (2019). Progressive delays in the timing of sardine migration in the southwest Indian Ocean. *South African Journal of Science, 115*(7/8), 1–6.

Fitchett, J.M., Grab, S.W., & Thompson, D.I. (2015). Plant phenology and climate change: Progress in methodological approaches and application. *Progress in Physical Geography, 39*(4), 460–482.

Fitchett, J.M., Grant, B., & Hoogendoorn, G. (2016a). Climate change threats to two low-lying South African coastal towns: Risks and perceptions. *South African Journal of Science, 112*(5–6), 1–9.

Fitchett, J.M., & Hoogendoorn, G. (2018). An analysis of factors affecting tourists' accounts of weather in South Africa. *International Journal of Biometeorology, 62*(12), 2161–2172.

Fitchett, J.M., & Hoogendoorn, G. (2019). Exploring the climate sensitivity of tourists to South Africa through TripAdvisor reviews. *South African Geographical Journal, 101*(1), 91–109.

Fitchett, J.M., Hoogendoorn, G., & Robinson, D. (2016b). Data challenges and solutions in the calculation of Tourism Climate Index (TCI) scores in South Africa. *Tourism: An International Interdisciplinary Journal, 64*(4), 359–370.

Fitchett, J.M., Hoogendoorn, G., & Swemmer, A.M. (2016c). Economic costs of the 2012 floods on tourism in the Mopani District Municipality, South Africa. *Transactions of the Royal Society of South Africa, 71*(2), 187–194.

Fitchett, J.M., Hoogendoorn, G., & van Tonder, S.M. (2021). Tropical cyclone and tourism: The case of the South West Indian Ocean. In P. Girish, & C.M. Hall (Eds.), *Aspects of tourism: Tourism, cyclones/hurricanes* (in press). Channelview Publications.Fitchett, J.M., Robinson, D., & Hoogendoorn, G. (2017). Climate suitability for tourism in South Africa. *Journal of Sustainable Tourism, 25*(6), 851–867.

Flato, G., Marotzke, J. Abiodun, B., Braconnot, P., Chou, S.C., Collins, W., Cox, P., Driouech, F., Emori, S., Eyring, V., Forest, C., Gleckler, P., Guilyardi, E., Jakob, C., Kattsov, V., Reason, C., & Rummukainen, M. (2013). Evaluation of climate models. In T.F. Stocker, D. Qin, G-K. Plattner, M. Tignor, S.K. Allen, J. Boschung, A. Nauels, Y. Xia, V. Bex, & P.M. Midgley (Eds.), *Climate change 2013: The physical science basis. Contribution of working group I to the Fifth Assessment Report of the Intergovernmental Panel on Climate Change.* Cambridge University Press.

Folke, C. (2006). Resilience: The emergence of a perspective for social-ecological systems analyses. *Global Environmental Change, 16*, 253–267.

Folland, C.K., Karl, T.R., & Salinger, M.J. (2002). Observed climate variability and change. *Weather, 57*(8), 269–278.

Ford, J.D., Berrang-Ford, L., Biesbroek, R., Araos, M., Austin, S.E., & Lesnikowski, A. (2015). Adaptation tracking for a post-2015 climate agreement. *Nature Climate Change, 5*, 967–969.

Franklin, A., & Crang, M. (2001). The trouble with tourism and travel theory? *Tourist Studies, 1*(1), 5–22.

Frey, N., & George, R. (2010). Responsible tourism management: The missing link between business owners' attitudes and behaviour in the Cape Town tourism industry. *Tourism Management, 31*(5), 621–628.

Friedrich, J., Stahl, J., Fitchett, J.M., & Hoogendoorn, G. (2020a). To beach or not to beach? Socio-economic factors influencing beach tourists' perceptions of climate and weather in South Africa. *Transactions of the Royal Society of South Africa, 75*(2), 194–202.

Friedrich, J., Stahl, J., Hoogendoorn, G., & Fitchett, J.M. (2020b). Exploring climate change threats to beach tourism destinations: Application of the hazard-activity pairs methodology to South Africa. *Weather, Climate, and Society, 12*(3), 529–544.

Friedrich, J., Stahl, J., Hoogendoorn, G., & Fitchett, J.M. (2021). Locational heterogeneity in climate change threats to beach tourism destinations in South Africa. In B. Lubbe, N. Moatshe, & J. Saarinen (Eds.), *Southern African perspectives on sustainable tourism development: Tourism and changing localities* (in press), Springer.

Gabarda-Mallorquí, A., Garcia, X., & Ribas, A. (2017). Mass tourism and water efficiency in the hotel industry: A case study. *International Journal of Hospitality Management, 61*, 82–93.

Gamage, S.K.N., Kuruppuge, R.H., & ul Haq, I. (2017). Energy consumption, tourism development, and environmental degradation in Sri Lanka. *Energy Sources, Part B: Economics, Planning and Policy, 12*(10), 910–916.

Garland, R.M., Matooane, M., Engelbrecht, F.A., Bopape, M.J.M., Landman, W.A., Naidoo, M., Merwe, J.V.D., & Wright, C.Y. (2015). Regional projections of extreme apparent temperature days in Africa and the related potential risk to human health. *International Journal of Environmental Research and Public Health, 12*(10), 12577–12604.

George, R. (2010). Visitor perceptions of crime-safety and attitudes towards risk: The case of Table Mountain National Park, Cape Town. *Tourism Management, 31*(6), 806–815.

Giddy, J.K. (2019). The impact of extreme weather on mass-participation sporting events. *International Journal of Event and Festival Management, 10*(2), 95–109.

Giddy, J.K., Fitchett, J.M., & Hoogendoorn, G. (2016). The influence of weather on South African tourism experiences: American perspectives. *Proceedings of the Centenary Conference of the Society of South African Geographers, Stellenbosch* (25–28 September 2016).

Giddy, J.K., Fitchett, J.M., & Hoogendoorn, G. (2017a). A case study into the preparedness of white-water tourism to severe climatic events in Southern Africa. *Tourism Review International, 21*(2), 213–220.

Giddy, J.K., Fitchett, J.M., & Hoogendoorn, G. (2017b). Insight into American tourists' experiences with weather in South Africa. *Bulletin of Geography: Socio-Economic Series, 38*, 57–72.

Gilbert, E.W. (1939). The growth of island and seaside health resorts in England. *Scottish Geographical Magazine, 55*(1), 16–35.

Gill, A.M., & Williams, P.W. (2011). Rethinking resort growth: Understanding evolving governance strategies in Whistler, British Columbia. *Journal of Sustainable Tourism, 19*(4–5), 629–648.

Gill, A.M., & Williams, P.W. (2014). Mindful deviation in creating a governance path towards sustainability in resort destinations. *Tourism Geographies, 16*(4), 546–562.

Gisselquist, R.M. (2012). Good governance as a concept, and why this matters for development policy. *WIDER Working Paper, 2012/30.*

Goldreich, Y. (2006). Ground and top canopy layer urban heat island partitioning on an airborne image. *Remote Sensing of Environment, 104*(2), 247–255.

Goliger, A.M., & Retief, J.V. (2007). Severe wind phenomena in southern Africa and the related damage. *Journal of Wind Engineering and Industrial Aerodynamics, 95*(9–11), 1065–1078.

Gómez Martin, M.B. (2005). Weather, climate and tourism – A geographical perspective. *Annals of Tourism Research, 32*(3), 571–591.

Gori, A., Lin, N., & Smith, J. (2020). Assessing compound flooding from landfalling tropical cyclones on the North Carolina Coast. *Water Resources Research, 56*(4), e2019WR026788.

Gössling, S., Bredberg, M., Randow, A., Sandström, E., & Svensson, P. (2006). Tourist perceptions of climate change: A study of international tourists in Zanzibar. *Current Issues in Tourism, 9*, 419–435.

Gössling, S., & Hall, C.M. (2006). Uncertainties in predicting tourist flows under scenarios of climate change. *Climatic Change, 79*(3), 163–173.

Gössling, S., Hall, C.M., & Scott, D. (2009). The challenges of tourism as a development strategy in an era of global climate change. In E. Palosuo (Ed.), *Rethinking development in a carbon-constrained world: Development cooperation and climate change* (pp. 100–109). Ministry of Foreign Affairs.

Gössling, S., Hall, C.M., & Scott, D. (2015). *Tourism and water.* Channelview Publications.

Gössling, S., & Peeters, P. (2007). 'It does not harm the environment!' An analysis of industry discourses on tourism, air travel and the environment. *Journal of Sustainable Tourism, 15*(4), 402–417.

Gössling, S., & Peeters, P. (2015). Assessing tourism's global environmental impact 1900–2050, *Journal of Sustainable Tourism, 23*(5), 639–659.Gössling, S., & Scott, D. (2018). The decarbonisation impasse: Global tourism leaders' views on climate change mitigation. *Journal of Sustainable Tourism, 26*(12), 2071–2086.

Gössling, S., Scott, D., & Hall, C.M. (2021). Pandemics, tourism and global change: A rapid assessment of COVID-19. *Journal of Sustainable Tourism, 29(1)*, 1–20.

Gössling, S., Scott, D., Hall, C.M., Ceron, J.P., & Dubois, G. (2012). Consumer behaviour and demand response of tourists to climate change. *Annals of Tourism Research, 39*(1), 36–58.

Göttsche, F.M., & Hulley, G.C. (2012). Validation of six satellite-retrieved land surface emissivity products over two land cover types in a hyper-arid region. *Remote Sensing of Environment, 124*, 149–158.

Government of Botswana. (2007). *Community-based natural resources management policy. Government paper 2.* Ministry of Environment, Wildlife and Tourism.

Government of Botswana. (2017). *National development plan 11.* Government Printer.

Government of the Republic of Namibia. (2011). *National policy on climate change for Namibia – 2011.* Government of the Republic of Namibia.

Grab, S., Linde, J., & De Lemos, H. (2017). Some attributes of snow occurrence and snowmelt/sublimation rates in the Lesotho Highlands: Environmental implications. *Water SA, 43*(2), 333–342.

Graciano, J.C., Ángeles, M., & Gámez, A.E. (2020). A critical geography approach to land and water use in the tourism economy in Los Cabos, Baja California Sur, Mexico. *Journal of Land Use Science, 15*(2–3), 439–456.

Green, I., & Saarinen, J. (2022). Changing environment and the political ecology of authenticity in heritage tourism: A case of the Ovahimba and the Ju/'Hoansi-San Living Museums in Namibia. In J. Saarinen, N. Moswete, & B. Lubbe (Eds.), *Southern African perspectives on sustainable tourism development: Tourism and changing localities* (in press). Springer.

Haddad, B.M. (2005). Ranking the adaptive capacity of nations to climate change when socio-political goals are explicit. *Global Environmental Change, 15*(2), 165–176.

Hall, C.M. (2000). *Tourism planning: Policies, processes and relationships*. Pearson.

Hall, C.M. (2009). Degrowing tourism: Décroissance, sustainable consumption and steady-state tourism. *Anatolia, 20*(1), 46–61.

Hall, C.M. (2011). Policy learning and policy failure in sustainable tourism governance: From first- and second-order to third-order change? *Journal of Sustainable Tourism, 19*(4–5), 649–671.

Hall, C.M. (2016). Heritage, heritage tourism and climate change. *Journal of Heritage Tourism, 11*(1), 1–9.

Hall, C.M., & Higham, J. (Eds.) (2005). *Tourism, recreation and climate change*. Channelview Publications.

Hall, C.M., Le-Klähn, D-T., & Ram, Y. (2017). *Tourism, public transport and sustainable mobility*. Channel View Publications.

Hall, C.M., Malinen, S., Vosslamber, R., & Wordsworth, R. (2016). *Business and post-disaster management: Business, organisational and consumer resilience and the Christchurch earthquakes*. Routledge.

Hall, C.M., Prayang, G., & Amore, A. (2018). *Tourism and resilience. Individual, organizational and destination perspectives*. Channel View Publications.

Hall, C.M., Scott, D., & Gössling, S. (2013). The primacy of climate change for sustainable international tourism. *Sustainable Development, 21*(2), 112–121.

Hall, C.M., Scott, D., & Gössling, S. (2020). Pandemics, transformations and tourism: be careful what you wish for. *Tourism Geographies, 22*(3), 577–98.

Hallegette, S. (2009). Strategies to adapt to an uncertain climate change. *Global Environmental Change, 19*, 240–247.

Hambira, W. (2011). Screening for climate change vulnerability in Botswana's tourism sector in a bid to explore suitable adaptation measures and policy implications: A case study of the Okavango Delta. *International Journal of Tourism Policy, 4*(1), 51–65.

Hambira, W. (2018). Botswana tourism operators' and policy makers' perceptions and responses to climate change. *Matkailututkimus, 14*(1), 55–59.

Hambira, W.L. (2017). Botswana tourism operators and policy makers' perceptions and responses to the tourism-climate change nexus: Vulnerabilities and adaptations to climate change in Maun and Tshabong area. *Nordia Geographical Publications, 46*(2), 1–59.

Hambira, W.L., Manwa, H., Atlhopheng, J., & Saarinen, J. (2015). Perceptions of tourism operators towards adaptations to climate change in nature-based tourism: The quest for sustainable tourism in Botswana. *Pula: Botswana Journal of African Studies, 27*(1), 69–85.

Hambira, W.L., & Mbaiwa, J.E. (2021). Tourism and climate change in Africa. In M. Novelli, E.A. Adu-Ampong, & M.A. Ribeiro (Eds.), *Routledge handbook of tourism in Africa* (pp. 126–145). Routledge.

Hambira, W.L., & Saarinen, J. (2015). Policy-makers' perceptions of the tourism-climate change nexus: Policy needs and constraints in Botswana. *Development Southern Africa, 32*(3), 350–362.

Hambira, W.L., Saarinen, J., Atlhopheng, J., & Manwa, H. (2021). Climate change, tourism and community development: Perceptions of Maun residents, Botswana. *Tourism Review International, 25*(2–3), 105–117.

Hambira, W.L., Saarinen, J., Manwa, H., & Atlhopeng, J. (2013). Climate change adaption practices in nature-based tourism in Maun in the Okavango Delta Area, Botswana: How prepared are the tourism businesses? *Tourism Review International, 17*(2), 19–29.

Hambira, W.L., Saarinen, J., & Moses, O. (2020). Climate change policy in a world of uncertainty: Changing environment, knowledge, and tourism in Botswana. *African Geographical Review, 39*(3), 252–266.

Hamilton, J., & Lau M. (2005). The role of climate information in tourist destination choice decision-making. In S. Gössling, & C.M. Hall (Eds.), *Tourism, recreation and climate change* (pp. 229–250). Routledge.

Hamilton, J., Maddison, D., & Tol, R. (2005). Climate change and international tourism: A simulation study. *Global Environmental Change, 15*, 253–266.

Hannam, K., & Butler, G. (2012). Engaging the new mobilities paradigm in contemporary African tourism research. *Africa Insight, 42*(2), 127–135.

Hannam, K., Butler, G., & Paris, C. (2014). Development and key issues in tourism mobilities. *Annals of Tourism Research, 44*, 171–185.

Hänsler, A., Cermak, J., Hagemann, S., & Jacob, D. (2011). Will the southern African west coast fog be affected by future climate change? Results of an initial fog projection using a regional climate model. *Erdkunde, 65*, 261–275.

Harms, T.M. (2006). *Energy, entropy, sunny skies and South Africa.* Stellenbosch University Press.

Harris, L., Nel, R., & Schoeman, D. (2011). Mapping beach morphodynamics remotely: A novel application tested on South African sandy shores. *Estuarine, Coastal and Shelf Science, 92*(1), 78–89.

Hart, N.C.G., Reason, C.J.C., & Fauchereau, N. (2013). Cloud bands over southern Africa: Seasonality, contribution to rainfall variability and modulation by the MJO. *Climate Dynamics, 41*, 1199–1212.

Hart, N.G.C., Washington, R., & Reason, C.J.C. (2018). On the likelihood of tropical-extratropical cloud bands in the South Indian Convergence Zone during ENSO events. *Journal of Climate, 31*, 2797–2817.

Hejazizadeh, Z., Karbalaee, A., Hosseini, S.A., & Tabatabaei, S.A. (2019). Comparison of the holiday climate index (HCI) and the tourism climate index (TCI) in desert regions and Makran coasts of Iran. *Arabian Journal of Geosciences, 12*, 803.

Herbertson, A.J. (1901). *Outlines of physiography: An introduction to the study of the earth.* Edward Arnold.

Higgins-Desbiolles, F. (2020). Socialising tourism for social and ecological justice after COVID-19. *Tourism Geographies, 22*(3), 610–23.

Higham, J., Cohen, S.A., Cavaliere, C.T., Reis, A., & Finkler, W. (2016). Climate change, tourist air travel and radical emissions reduction. *Journal of Cleaner Production, 111*, 336–347.

Hinkel, J. (2011). Indicators of vulnerability and adaptive capacity: Towards a clarification of the science–policy interface. *Global Environmental Change, 21*(1), 198–208.

Hoegh-Guldberg, O., Jacob, D., Taylor, M., Bindi, M., Brown, S., Camilloni, I., Diedhiou, A. Djalante, R., Ebi, K.L., Engelbrecht, F., Guiot, J., Hijioka, Y., Mehrotra, S., Payne, A., Seneviratne, S.I., Thomas, A., & Zhou, G. (2018). Impacts of 1.5°C global warming on natural and human systems. In V. Masson-Delmotte, P. Zhai, H.-O. Pörtner, D. Roberts, J. Skea, P.R. Shukla, A. Pirani, W. Moufouma-Okia, C. Péan, R. Pidcock, S. Connors, J.B.R. Matthews, Y. Chen, X. Zhou, M.I. Gomis, E. Lonnoy, T. Maycock, M. Tignor, & T. Waterfield (Eds.), *Global Warming of 1.5°C. An IPCC Special Report on the impacts of global warming of 1.5°C above pre-industrial levels and related global greenhouse gas emission pathways, in the context of strengthening the global response to the threat of climate change, sustainable development, and efforts to eradicate poverty* (pp. 175–311). IPCC.

Hollenhorst, S.J., Houge-MacKenzie, S., & Ostergren, D.M. (2014). The trouble with tourism. *Tourism Recreation Research, 39*(3), 305–319.

Holling, C.S. (1973). Resilience and stability of ecological systems. *Annual Review of Ecology and Systematics, 4*, 1–23. https://doi.org/10.1146/annurev.es.04.110173.000245

Hoogendoorn, G. (2011a). Low income earners as second home tourists in South Africa? *Tourism Review International, 15*(1–2), 37–50.

Hoogendoorn, G. (2011b). *Second homes and local economic impacts in the South Africa post-productivist countryside.* PhD Thesis submitted to the University of the Free State, Bloemfontein.

Hoogendoorn, G. (2014). Mapping fly-fishing tourism in Southern Africa. *African Journal of Hospitality, Tourism and Leisure, 3*(2), 1–13.

Hoogendoorn, G. (2021). Last chance tourism in South Africa: Future research potential? *Tourism: An International Interdisciplinary Journal, 69*(1), 73–82.

Hoogendoorn, G., & Back, A. (2019). Snowballing in 35°C: An inquiry into second-home tourism in Mozambique. *Tourism: An International Interdisciplinary Journal, 67*(3), 311–317.

Hoogendoorn, G., & Fitchett, J.M. (2018a). Tourism and climate change: A review of threats and adaptation strategies for Africa. *Current Issues in Tourism, 21*(7), 742–759.

Hoogendoorn, G., & Fitchett, J.M. (2018b). Perspectives on second homes, climate change and tourism in South Africa. *African Journal of Hospitality, Tourism and Leisure, 7*(2), 1–18.

Hoogendoorn, G., & Fitchett, J.M. (2020). Fourteen years of tourism and climate change research in Southern Africa. In M.T. Stone, M. Lenao, & M. Moswete (Eds.), *Natural resources, tourism and community livelihoods in southern Africa: Challenges of Sustainable Development* (pp. 78–89). Routledge.

Hoogendoorn, G., & Giddy, J.K. (2017). "Does this look like a slum?" Walking tours in the Johannesburg inner city. *Urban Forum, 28*(3), 315–328.

Hoogendoorn, G., Grant, B., & Fitchett, J. (2015). Towards green guest houses in South Africa: The case of Gauteng and KwaZulu-Natal. *South African Geographical Journal, 97*(2), 123–138.

Hoogendoorn, G., Grant, B., & Fitchett, J.M. (2016). Disjunct perceptions? Climate change threats in two-low lying South African coastal towns. *Bulletin of Geography. Socio-Economic Series, 31*, 59–71.

Hoogendoorn, G., Kelso, C., & Sinthumule, I. (2019). Tourism in the Great Limpopo Transfrontier Park: A review. *African Journal of Hospitality, Tourism and Leisure, 8*(5), 1–15.

Hoogendoorn, G., & Rogerson, C.M. (2016). New perspectives on Southern African tourism research. *Tourism: An International Interdisciplinary Journal, 64*(4), 355–357.

Hoogendoorn, G., Stockigt, L., Saarinen, J., & Fitchett, J.M. (2021). Adapting to climate change: The case of snow-based tourism in Afriski, Lesotho. *African Geographical Review, 40*(1), 92–104.

Hoogendoorn, G., & Visser, G. (2011). Tourism, second homes, and emerging South African post-productivist countryside. *Tourism Review International, 15*(1/2), 183–197.

Hoogendoorn, G., & Visser, G. (2012). Stumbling over researcher positionality and political-temporal contingency in South African second-home tourism research. *Critical Arts, 26*(3), 254–271.

Howard, E., & Washington, R. (2019). Drylines in southern Africa: Rediscovering the congo air boundary. *Journal of Climate, 32*(23), 8223–8242.

Hu, Y., & Ritchie, J.R.B. (1993). Measuring destination attractiveness: A contextual approach. *Journal of Travel Research, 32*(2), 25–34.

Huang, Z., Cao, F., Jin C., Yu, Z., & Huang, R. (2017). Carbon emission flow from self-driving tours and its spatial relationship with scenic spots – A traffic-related big data method. *Journal of Cleaner Production, 142*, 946–955.

Hughes, E., & Scheyvens, R. (2016). Corporate social responsibility in tourism post-2015: A development first approach. *Tourism Geographies, 18*(5), 469–482.

Hyam, R., & Henshaw, P. (2003). *The Lion and the Springbok: Britain and South Africa since the Boer War.* Cambridge University Press.

Hyman, T-A. (2014). Assessing the vulnerability of beach tourism and non-beach tourism to climate change: A case study from Jamaica. *Journal of Sustainable Tourism, 22*(8), 1197–1215.

IPCC (Intergovernmental Panel on Climate Change). (2007). Climate change 2007: Working group II: Impacts, adaptation and vulnerability. Summary for policy makers. IPCC. https://www.ipcc.ch/publications_and_data/ar4/wg2/en/spmsspm-e.html

IPCC (Intergovernmental Panel on Climate Change). (2021). *Climate change 2021: The physical science basis. Contribution of working group I to the sixth assessment report of the intergovernmental panel on climate change.* Cambridge University Press.

Islam, W., Ruhanen, L., & Ritchie, B.W. (2018). Tourism governance in protected areas: Investigating the application of the adaptive co-management approach. *Journal of Sustainable Tourism, 18*(11), 1890–1908.

Jamal, T., & Camargo, B.A. (2014). Sustainable tourism, justice and an ethic of care: Toward the just destination. *Journal of Sustainable Tourism, 22*(1), 11–30.

Jamal, T., & Camargo, B.A. (2018). Tourism governance and policy: Whither justice? *Tourism Management Perspectives, 25*, 205–208.

James, R., Hart, N., Munday, C., Reason, C., & Washington, R. (2020). Coupled climate model simulations of Tropical-Extratropical cloud bands over southern Africa. *Journal of Climate, 33*(19), 8579–8602.

James, R., & Washington, R. (2013). Changes in African temperature and precipitation associated with degrees of global warming. *Climatic Change, 117*(4), 859–872.

James, R., Washington, R., Abiodun, B., Kay, G., Mutemi, J., Pokam, W., Hart, N., Artan, G., & Senior, C. (2018). Evaluating climate models with an African lens. *Bulletin of the American Meteorological Society, 99*(2), 313–336.

Jeacle, I., & Carter, C. (2011). In TRIPADVISOR we trust: Rankings, calculative regimes and abstract systems. *Accounting, Organizations and Society, 36*, 293–309.

Jelle, B.P. (2011). Traditional, state-of-the-art and future thermal building insulation materials and solutions – Properties, requirements and possibilities. *Energy and Buildings, 43*(10), 2549–2563.

Jeuring, J.H.G. (2017). Weather perceptions, holidays satisfaction and perceived attractiveness of domestic vacationing in The Netherlands. *Tourism Management, 61*, 70–81.

Jiménez-Aceituno, A., Peterson, G.D., Norström, A.V., Wong, G.Y., & Downing, A.S. (2020). Local lens for SDG implementation: Lessons from bottom-up approaches in Africa. *Sustainability Science, 15*, 729–743.

Jin, C., Cheng, J., Xu, J., & Huang, Z. (2018). Self-driving tourism induced carbon emission flows and its determinants in well-developed regions: A case study of Jiangsu Province, China. *Journal of Cleaner Production, 186*, 191–202.

Jordhus-Lier, D., Saaghus, A., Scott, D., & Ziervogel, G. (2019). Adaptation to flooding, pathway to housing or 'wasteful expenditure'? Governance configurations and local policy subversion in a flood-prone informal settlement in Cape Town. *Geoforum, 98*, 55–65.

Joshi, M., Hawkins, E., Sutton, R., Lowe, J., & Frame, D. (2011). Projections of when temperature change will exceed 2°C above pre-industrial levels. *Nature Climate Change, 1*(8), 407–412.

Jourdan, D., & Wertin, J. (2020). Intergenerational rights to a sustainable future: Insights for climate justice and tourism. *Journal of Sustainable Tourism, 28*(8), 1245–1254.

Jury, M.R., Valentine, H.R., & Lutjeharms, J.R.E. (1993). Influence of the Agulhas current on summer rainfall along the south-east coast of South Africa. *Journal of Applied Meteorology, 32*, 1282–1287.

Jury, M.R., White, W.B., & Reason, C.J.C. (2004). Modelling the dominant climate signals around southern Africa. *Climate Dynamics, 23*, 717–726.

Kabat, P., van Vierssen, W., Veraart, J., Vellinga, P., & Aerts, J. (2005). Climate proofing the Netherlands. *Nature, 438*, 283–284.

Kaján, E., & Saarinen, J. (2013). Tourism, climate change and adaptation: A review. *Current Issues in Tourism, 16*(2), 167–195.

Kaján, E., Tervo-Kankare, K., & Saarinen, J. (2015). Cost of adaptation to climate change in tourism: Methodological challenges and trends for future studies in adaptation. *Scandinavian Journal of Hospitality and Tourism, 15*(3), 311–317.

Karthikeyan, T., Sekaran, K., Ranjith, D., & Balajee, J. M. (2019). Personalized content extraction and text classification using effective web scraping techniques. *International Journal of Web Portals, 11*(2), 41–52.

Kasirye, I., Ntlale, A., & Venugopal, G. (2020). Implementation progress of the SDGs: Sub-Saharan Africa regional survey. *Southern Voice, Occasional Papers Series*, 66. http://southernvoice.org/wp-content/uploads/2020/07/Progress-SDGs-SSA-Kasirye-Ntla-Venugopal-2020.pdf

Kavita, E., & Saarinen, J. (2016). Tourism and rural community development in Namibia: Policy issues review. *Fennia, 194*(1), 79–88.

Keja-Kaereho, C., & Tjizu, B. (2019). Climate change and global warming in Namibia: Environmental disasters vs human life and the economy. *Management and Economics Research Journal, 5*(1), 1–11.

Khan, M.A., Khan, M.Z., Zaman, K., & Naz, L. (2014). Global estimates of energy consumption and greenhouse gas emissions. *Renewable and Sustainable Energy Reviews, 29*, 336–344.

Khazai, B., Mahdavian, F., & Platt, S. (2018). Tourism Recovery Scorecard (TOURS) – Benchmarking and monitoring progress on disaster recovery in tourism destinations. *International Journal of Disaster Risk Reduction, 27*, 75–84.

Khumalo, T., Sebatlelo, P., & van der Merwe, C.D. (2014). Who is a heritage tourist? A comparative study of Constitution Hill and the Hector Pieterson Memorial and Museum, Johannesburg, South Africa. *African Journal of Hospitality, Tourism and Leisure, 3*(1), 1–13.

Klein, R.A. (2010). The cruise sector and its environmental impact. In C. Schott (Ed.), *Tourism and the implications of climate change: Issues and actions* (pp. 113–130). Emerald Group Publishing Limited.

Klotzbach, P.J., Chan, J.C., Fitzpatrick, P.J., Frank, W.M., Landsea, C.W., & McBride, J.L. (2017). The science of William M. Gray: His contributions to the knowledge of tropical meteorology and tropical cyclones. *Bulletin of the American Meteorological Society, 98*(11), 2311–2336.

König, U., & Abegg, B. (1997). Impacts of climate change on winter tourism in the Swiss Alps. *Journal of Sustainable Tourism, 5*(1), 46–58.

Köppen, W. (1936). Das geographisca system der klimate. In W. Köppen, & G. Geiger (Eds.), *Handbuch der klimatologie* (p. 4). Gerb.

Kottek, M., Grieser, J., Beck, C., Rudolf, B., & Rubel, F. (2006). World map of the Köppen-Geiger climate classification updated. *Meteorologische Zeitschrift, 15*(3), 259–263.

Kourtit, K., Nijkamp, P., & Romão, J. (2019). Cultural heritage appraisal by visitors to global cities: The use of social media and urban analytics in urban buzz research. *Sustainability, 11*(3470), 3470.

Kraaij, T., Baard, J.A., Arndt, J., Vhengani, L., & van Wilgen, B.W. (2018). An assessment of climate, weather and fuel factors influencing a large, destructive wildfire in the Knysna region, South Africa. *Fire Ecology, 14*(4), 1–12.

Kruger, A.C. (2006). Observed trends in daily precipitation indices in South Africa: 1910–2004. *International Journal of Climatology, 26*, 2275–2285.

Kruger, A.C., Goliger, A.M., Retief, J.V., & Sekele, S. (2010). Strong wind climatic zones in South Africa. *Wind and Structures, 13*(1), 37–55.

Kruger, A.C., & Sekele, S.S. (2013). Trends in extreme temperature indices in South Africa: 1962–2009. *International Journal of Climatology, 33*, 661–676.

Kruger, A.C., & Shongwe, S. (2004). Temperature trends in South Africa: 1960–2003. *International Journal of Climatology, 24*, 1929–1945.

Kusangaya, S., Warburton, M.L., Van Garderen, E.A., & Jewitt, G.P. (2014). Impacts of climate change on water resources in southern Africa: A review. *Physics and Chemistry of the Earth, 67*, 47–54.

Kutzner, D. (2019). Environmental change, resilience, and adaptation in nature-based tourism: Conceptualizing the social-ecological resilience of birdwatching tour operations. *Journal of Sustainable Tourism, 27*(8), 1142–1166.

Lai, P.H., Hsu, Y.C., & Wearing, S. (2016). A social representation approach to facilitating adaptive co-management in mountain destinations managed for conservation and recreation. *Journal of Sustainable Tourism, 24*(2), 227–244.

Landers, R.N., Brusso, R.C., Cavanaugh, K.J., & Collmus, A.B. (2016). A primer on theory-driven web scraping: Automatic extraction of big data from the Internet for use in psychological research. *Psychological Methods, 21*(4), 475.

Lapeyre, R. (2010). Community-based tourism as a sustainable solution to maximise impacts locally? The Tsiseb Conservancy case, Namibia. *Development Southern Africa, 27*, 757–772.

Lapeyre, R. (2011). The tourism global commodity chain in Namibia: Industry concentration and its impacts on transformation. *Tourism Review International, 15*(1–2), 63–75.

Latour, B. (2015). Telling friends from foes in the time of the Anthropocene. In C. Hamilton, F. Gemenne, & C. Bonneuil (Eds.), *The Anthropocene and the global environmental crisis: Rethinking modernity in a new epoch* (pp. 145–155). Routledge.

Lavers, J.L., Rivers-Auty, J., & Bond, A.L. (2021). Plastic debris increases circadian temperature extremes in beach sediments. *Journal of Hazardous Materials, 416*, 126140.

Lee, J.Y., Marotzke, J., Bala, G., Cao, L., Corti, S., Dunne, J.P., Engelbrecht, F., Fischer, E., Fyfe, J.C., Jones, C., Maycock, A., Mutemi, J., Ndiaye, O., Panickal, S., &

Zhou, T. (2021). Future global climate: Scenario-based projections and near-term information. In V. Masson-Delmotte, P. Zhai, A. Pirani, S. L. Connors, C. Péan, S. Berger, N. Caud, Y. Chen, L. Goldfarb, M. I. Gomis, M. Huang, K. Leitzell, E. Lonnoy, J. B. R. Matthews, T. K. Maycock, T. Waterfield, O. Yelekçi, R. Yu, & B. Zhou (Eds.), *Climate change 2021: The physical science basis. Contribution of working group I to the sixth assessment report of the intergovernmental panel on climate change.* Cambridge University Press (in press).

Lee, S., & Jamal, T. (2008). Environmental justice and environmental equity in tourism: Missing links to sustainability. *Journal of Ecotourism, 7*(1), 44–67.

Lehmann, D., Mfune, J.K E., Gewers, E., Cloete, J., Aschenborn, O.K., Mbomboro, L., Kasaona, S., Brain, C., & Voigt, C.C. (2020). Spatiotemporal responses of a desert dwelling ungulate to increasing aridity in North-eastern Namibia. *Journal of Arid Environments, 179*, 104193.

Leiper, N. (1979). The framework of tourism. *Annals of Tourism Research, 6*, 390–407.

Lenao, M., & Basupi, B. (2016). Ecotourism development and female empowerment in Botswana: A review. *Tourism Management Perspectives, 18*, 51–58.

Lennard, C. (2019). Multi-scale drivers of the South African weather and climate. In J. Knight, & C.M. Rogerson (Eds.), *The geography of South Africa: Contemporary changes and new directions* (pp. 81–89). Springer.

Lenzen, M., Sun, Y-Y., Faturay, F., Ting, Y-P., Geschke, A., & Malik, A. (2018). The carbon footprint of global tourism. *Nature Climate Change, 8*, 522–528.

Lesolle, D., & Ndzinge, I.N. (2018). *UNDP project: Environment and climate change contract no. INST 001/2017.* United Nations Development Programme.

Lew, A.A. (2014). Scale, change and resilience in community tourism planning. *Tourism Geographies, 16*(1), 14–22.

Lew, A.A., & Cheer, J. (2018). Environmental change, resilience and tourism. In A.A. Lew, & J. Cheer (Eds.), *Tourism resilience and adaptation to environmental change: Definitions and frameworks* (pp. 3–12). Routledge.

Lew, A.A., Ng, P.T., Ni, C., & Wu, T. (2016). Community sustainability and resilience: Similarities, differences and indicators. *Tourism Geographies, 18*(1), 18–27.

Liang, X., & Chan, J.C. (2005). The effects of the full Coriolis force on the structure and motion of a tropical cyclone. Part I: Effects due to vertical motion. *Journal of the Atmospheric Sciences, 62*(10), 3825–3830.

Liburd, J., Duedahl, E., & Heape, C. (2020). Co-designing tourism for sustainable development. *Journal of Sustainable Tourism*, 1–20. https://doi.org/10.1080/09669 582.2020.1839473

Lin, T., & Matzarakis, A. (2011). Tourism climate information based on human thermal perception in Taiwan and Eastern China. *Tourism Management, 32*(3), 492–500.

Lise, W., & Tol, R.S.J. (2002). Impact of climate on tourism demand. *Climatic Change, 55*(4), 429–449.

Little, M.E., & Blau, E. (2020). Social adaptation and climate mitigation through agrotourism: A case study of tourism in Mastatal, Costa Rica. *Journal of Ecotourism, 19*(2), 97–112.

Lohmann, M., & Kaim, E. (1999). Weather and holiday destination preferences image, attitude and experience. *The Tourist Review, 54*(2), 54–64.

López-Bonilla, L.M., Reyes-Rodríguez, C., & López-Bonilla, J.M. (2020). Golf tourism and sustainability: Content analysis and directions for future research. *Sustainability, 12*(9), 3616.

Lukhele, S.E., & Mearns, K.F. (2013). The operational challenges of community-based tourism ventures in Swaziland. *African Journal of Physical, Health Education, Recreation and Dance, 2*, 199–216.

Luthje, T., & Wyss, R. (2014). Assessing and planning resilience in tourism. *Tourism Management, 44*, 161–163.

Ma, S., Craig, C.A., & Feng, S. (2020). The Camping Climate Index (CCI): The development, validation, and application of a camping-sector tourism climate index. *Tourism Management, 80*, 104105.

Machete, F., Hongoro, C., Nhamo, G., & Mearns, K. (2015). Influence of energy saving on the quality of lighting services on selected hotels in Mpumalanga, Republic of South Africa. *African Journal of Science, Technology, Innovation and Development, 7*(4), 301–305.

Machete, F., Nhamo, G., & Rampedi, I. (2016). The association between energy saving and guest comfort in selected guesthouses in Mpumalanga Province, South Africa. *Africa Insight, 45*(4), 195–208.

Mackellar, N., New, M., & Jack, C. (2014). Observed and modelled trends in rainfall and temperature for South Africa: 1960–2010. *South African Journal of Science, 110*(7/8), 1–13.

Maddison, D. (2001). In search of warmer climates? The impact of climate change on flows of British tourists. *Climatic Change, 49*, 193–208.

Mahadew, R., & Appadoo, K.A. (2019). Tourism and climate change in Mauritius: Assessing the adaptation and mitigation plans and prospects. *Tourism Review, 74*(2), 204–215.

Mahlalela, P.T., Blamey, R.C., Hart, N.C.G., & Reason, C.J.C. (2020). Drought in the eastern Cape region of South Africa and trends in rainfall characteristics. *Climate Dynamics, 55*, 2743–2759.

Mahlangu, M.N., & Fitchett, J.M. (2019). Climate change threats to a floral wedding: Threats of shifting phenology to the emerging South African wedding industry. *Bulletin of Geography. Socio-Economic Series, 45*, 7–23.

Malan, N. (2018). Are icebergs a realistic option for augmenting Cape Town's water supply? *The Water Wheel, 17*(2), 32–34.

Malherbe, J., Engelbrecht, F.A., Landman, W.A., & Engelbrecht, C.J. (2012). Tropical systems from the southwest Indian Ocean making landfall over the Limpopo River Basin, southern Africa: A historical perspective. *International Journal of Climatology, 32*, 1018–1032.

Malherbe, J., Smit, I.P.J., Wessels, K.J., & Beukes, P.J. (2020). Recent droughts in the Kruger National Park as reflected in the extreme climate index. *African Journal of Range and Forage Science, 37*(1), 1–17.

Manatsa, D., & Reason, C. (2017). ENSO-Kalahari Desert linkages on southern African summer surface air temperature variability. *International Journal of Climatology, 37*, 1728–1745.

Manrai, L.A., Lascu, D-N., & Manrai, A.K. (2020). A study of safari tourism in sub-Saharan Africa: An empirical test of Tourism A-B-C (T-ABC) model. *Journal of Business Research, 119*, 639–651.

Manwa, H., Saarinen, J., Athlopheng, J., & Hambira, W.L. (2017). Sustainability management and tourism impacts on communities: Residents' attitudes in Maun and Tshabong, Botswana. *African Journal of Hospitality, Tourism and Leisure, 6*(3), 1–15.

Marshall, N.A., Marshall, P.A., Abdulla, A., Rouphael, T., & Ali, A. (2011). Preparing for climate change: Recognising its early impacts through the perceptions of dive tourists and dive operators in the Egyptian Red Sea. *Current Issues in Tourism, 14*(6), 507–518.

Marson, M., Maggi, E., & Scacchi, M. (2021). Financing African infrastructure: The role of China in African railways. *Research in Transportation Economics, 88*, 101111.

Martin, C.J. (2016). The sharing economy: A pathway to sustainability or a nightmarish form of neoliberal capitalism? *Ecological Economics, 121*, 149–159.

Martin, R., & Sunley, P. (2014). On the notion of regional economic resilience: Conceptualization and explanation. *Journal of Economic Geography, 15*(1), 1–42.

Martin, R., Sunley, P., Gardiner, B., & Tyler, P. (2016). How regions react to recessions: Resilience and the role of economic structure. *Regional Studies, 50*(4), 561–585.

Mason, S.J., & Joubert, A.M. (1997). Simulated changes in extreme rainfall over southern Africa. *International Journal of Climatology, 17*(3), 291–301.

Mather, A.A. (2007). Linear and nonlinear sea-level changes at Durban, South Africa. *South African Journal of Science, 103*, 509–512.

Mather, A.A., & Stretch, D.D. (2012). A perspective on sea level rise and coastal storm surge from Southern and Eastern Africa: A case study near Durban, South Africa. *Water, 4*(1), 237–259.

Mathivha, F.I., Tshipala, N.N., & Nkuna, Z. (2017). The relationship between drought and tourist arrivals: A case study of Kruger National Park, South Africa. *Jàmbá: Journal of Disaster Risk Studies, 9*(1), 1–8.

Matikinca, P., Ziervogel, G., & Enqvist, J.P. (2020). Drought response impacts on household water use practices in Cape Town, South Africa. *Water Policy, 22*(3), 483–500.

Matthews, L., Scott, D., Andrey, J., Mahon, R., Trotman, A., Burrowes, R., & Charles, A. (2021). Developing climate services for Caribbean tourism: A comparative analysis of climate push and pull influences using climate indices. *Current Issues in Tourism, 24*(11), 1576–1594.

Maunder, I. (1970). *The value of the weather*. Methuen.

Mauritius Meteorological Services. (2021, August 18). *Climate change*. http://metservice.intnet.mu/climate-services/climate-change.php

Mbaiwa, J., & Tshamekang, T. (2012). Developing a viable community-based tourism project in Botswana: The case of the Chobe Enclave conservation Trust. *World Sustainable Development Outlook*, 519–536.

Mbaiwa, J.E. (2005). Enclave tourism and its socio-economic impacts in the Okavango Delta, Botswana. *Tourism Management, 26*(2), 157–172.

Mbasera, M., Du Plessis, E., Saayman, M., & Kruger, M. (2016). Environmentally-friendly practices in hotels. *Acta Commercii-Independent Research Journal in the Management Sciences, 16*(1), 1–8.

Mbokodo, I., Bopape, M.J., Chikoore, H., Engelbrecht, F., & Nethengwe, N. (2020). Heatwaves in the future warmer climate of South Africa. *Atmosphere, 11*(7), 712.

McCarthy, J., Canziani, O.F., Leary, N.A., Dokken, D.J., & White, K.S. (Eds.) (2001). *Climate change 2001: Impacts, adaptation and vulnerability*. Cambridge University Press.

McCool, S. (2015). Sustainable tourism: Guiding fiction, social trap or path to resilience? In T.V. Singh (Ed.), *Challenges in tourism research* (pp. 224–234). Channel View Publications.

McKenna, A. (2010). *The history of southern Africa*. Britannica Education Publishing.

McKercher, B., Prideaux, B., Cheung, C., & Law, R. (2010). Achieving voluntary reductions in the carbon footprint of tourism and climate change. *Journal of Sustainable Tourism, 18*(3), 297–317.

McMurray, K.C. (1930). The use of land for recreation. *Annals of Association of the American Geographers, 20*(1), 7–20.

McPherson, F. (2020). *Climate change and heritage tourism: Threats to Makgabeng in a regional context, Limpopo South Africa.* MSc Dissertation submitted to the University of the Witwatersrand, Johannesburg.

Mearns, K.F., & Boshoff, D. (2017). Utilising sustainable tourism indicators to determine the environmental performance of the Sun City Resort, South Africa. *African Journal for Physical Activity and Health Sciences, 23*(1), 89–114.

Melubo, K. (2020). Is there room for domestic tourism in Africa? The case of Tanzania. *Journal of Ecotourism, 19*(3), 248–265.

Mendelsohn, R., Emanuel, K., Chonabayashi, S., & Bakkensen, L. (2012). The impact of climate change on global tropical cyclone damage. *Nature Climate Change, 2*(3), 205–209.

Meng, W., Xu, L., Hu, B., Zhou, J., & Wang, Z. (2016). Quantifying direct and indirect carbon dioxide emissions of the Chinese tourism industry. *Journal of Cleaner Production, 126*, 586–594.

Meredith, M. (2006). *The state of Africa: A history of fifty years of independence.* Free Press.

Mieczkowski, Z. (1985). The tourism climatic index: A method of evaluating world climate for tourism. *Canadian Geographer, 29*(3), 220–233

Milford, R.V., Goliger, A.M., & Vinsen, J.F. (1994). Tornado activity in South Africa. *Journal of the South African Institution of Civil Engineers, 36*(1), 17–23.

Millington, N., & Scheba, S. (2020). Day zero and the infrastructures of climate change: Water governance, inequality, and infrastructural politics in Cape Town's water crisis. *International Journal of Urban and Regional Research, 45*(1), 116–132.

Ministry Minerals, Energy and Water Resources. (2012). *Botswana national water policy.* Government Printer.

Ministry of Environment, Natural Resources Conservation and Tourism. (2021a). *Climate change policy.* Government Printer.

Ministry of Environment, Natural Resources Conservation and Tourism. (2021b). *The Botswana climate change policy.* The Government of Botswana.

Ministry of Environment and Tourism. (2018). *Tourist statistical report 2018.* Government of Namibia.

Mitchell, D., Henschel, J.R., Hetem, R.S., Wassenaar, T.D., Strauss, W.M., Hanrahan, S.A., & Seely, M.K. (2020). Fog and fauna of the Namib Desert: Past and future. *Ecosphere, 11*(1), e02996.

Morgan, R., Gatell, E., Junyent, R., Micallef, A., Özhan, E., & Williams, A.T. (2000). An improved user-based beach climate index. *Journal of Coastal Conservation, 6*(1), 41–50.

Morrissey, J., Moore, T., & Horne, R.E. (2011). Affordable passive solar design in a temperate climate: An experiment in residential building orientation. *Renewable Energy, 36*, 568–577.

Morton, T. (2012). From modernity to the Anthropocene: Ecology and art in the age of asymmetry. *International Social Science Journal, 63*(207–208), 39–51.

Moswete, N., & Dube, O.P. (2013). Wildlife-based tourism and climate: Potential opportunities and challenges for Botswana. In L. D'Amore & P. Kalifungwa

(Eds.), *Meeting the challenges of climate change to tourism: Case studies of best practice* (pp. 395–416). Cambridge Scholars.

Moswete, N., & Thapa, B. (2018). Local communities, CBOs/trusts, and people–park relationships: A case study of the Kgalagadi Transfrontier Park, Botswana. *The George Wright Forum, 35*(1), 96–108.

Mphale, K., Adedoyin, A., Nkoni, G., Ramaphane, G., Wiston, M., & Chimidza, O. (2018). Analysis of temperature data over semi-arid Botswana: Trends and break points. *Meteorology and Atmospheric Physics, 130*, 701–724.

Mukheibir, P., & Ziervogel, G. (2007). Developing a municipal adaptation plan (MAP) for climate change: The city of Cape Town. *Environment and Urbanization, 19*(1), 143–158.

Muller, M. (2018). Cape Town's drought: Don't blame climate change. *Nature, 559*, 174–176.

Murphy, C., Bastola, S., Hall, J., Harrigan, S., Murphy, N., & Holman, C. (2011). Against a 'wait and see' approach in adapting to climate change. *Irish Geography, 44*(1), 81–95.

Musavengane, R., Siakwah, P., & Leonard, L. (2020). The nexus between tourism and urban risk: Towards inclusive, safe, resilient and sustainable outdoor tourism in African cities. *Journal of Outdoor Recreation and Tourism, 29*, 100254.

Mushawemhuka, W., Fitchett, J.M., & Hoogendoorn, G. (2020). Towards quantifying climate suitability for Zimbabwean nature-based tourism. *South African Geographical Journal.* https://doi.org/10.1080/03736245.2020.1835703

Mushawemhuka, W., Fitchett, J.M., & Hoogendoorn, G. (2021). The influence of the media on perceptions of tourism and climate change at the Victoria Falls, Zimbabwe. *Weather, Climate and Society.* https://doi.org/10.1175/WCAS-D-21-0013.1

Mushawemhuka, W., Rogerson, J.M., & Saarinen, J. (2018). Nature-based tourism operators' perceptions and adaptation to climate change in Hwange National Park, Zimbabwe. *Bulletin of Geography. Socio-Economic Series, 42*, 115–127.

Mushawemhuka, W.J. (2021). *A comprehensive assessment of climate change threats and adaptation of nature-based tourism in Zimbabwe.* PhD thesis submitted to the University of Johannesburg.

Mwangi, O. (2007). Hydropolitics, ecocide and human security in Lesotho: A case study of the Lesotho Highlands Water Project. *Journal of Southern African Studies, 33*(1), 3–17.

Mycoo, M. (2014). Sustainable tourism, climate change and sea level rise adaptation policies in Barbados. *Natural Resources Forum, 38*, 47–57.

Naidu, N. (2008). Creating an African tourist experience at the cradle of humankind world heritage site. *Historia, 53*(2), 182–207.

Narita, D., Tol, R.S., & Anthoff, D. (2009). Damage costs of climate change through intensification of tropical cyclone activities: An application of FUND. *Climate Research, 39*(2), 87–97.

Navarro-Drazich, D. (2019). Climate change and tourism in Latin America. In C. Lorenzo (Ed.), *Latin America in Time of Global Environmental Change* (pp. 93–105). Springer.

Nel, P. (2003). Income inequality, economic growth, and political instability in sub-Saharan Africa. *The Journal of Modern African Studies, 41*(4), 611–639.

Neumann, B., Vafeidis, A.T., Zimmermann, J., & Nicholls, R.J. (2015). Future coastal population growth and exposure to sea-level rise and coastal flooding-a global assessment. *PloS One, 10*(3), e0118571.

New, M., & Bosworth, B. (2018, October 10). *What does global warming mean for Botswana and Namibia?* Thompson Reuters. https://news.trust.org/item/20181010140849-iaq7c/

New, M., Hewitson, B., Stephenson, D.B., Tsiga, A., Kruger, A., Manhique, A., Gomez, B., Coelho, C.A., Masisi, D.N., Kululanga, E., Mbambalala, E., Adesina, F., Saleh, H., Kanyanga, J., Adosi, J., Bulane, L., Fortunata, L., Mdoka, M.L., & Lajoie, R. (2006). Evidence of trends in daily climate extremes over southern and west Africa. *Journal of Geophysical Research: Atmospheres, 111*, D14.

News24. (2015, November 15). Heatwave stops Zimbabwe plane from landing. https://www.news24.com/News24/heatwave-stops-zimbabwe-plane-from-landing-20151115

Ngoni, S., & Saarinen, J. (2021). Community perceptions on the benefits and challenges of Community-based natural resources management in Zimbabwe. *Development Southern Africa*. https://doi.org/10.1080/0376835X.2020.1796599

Nguyen, X.P., Hoang, A.T., Ölçer, A.I., & Huynh, T.T. (2021). Record decline in global CO_2 emissions prompted by COVID-19 pandemic and its implications on future climate change policies. *Energy Sources, Part A: Recovery, Utilization, and Environmental Effects*. https://doi.org/10.1080/15567036.2021.1879969

Nhamo, G., Dube, K., & Chikodzi, D. (2020). *Counting the cost of COVID-19 on the global tourism industry*. Springer.

Nicholls, R.J. (2011). Planning for the impacts of sea level rise. *Oceanography, 24*(2), 144–157.

Nicholls, S. (2014). *Climate change: Implications for tourism*. Cambridge Institute for Sustainability Leadership. Cambridge University.

Nilsson, J.H., & Gössling, S. (2013). Tourist responses to extreme environmental events: The case of Baltic Sea Algal Blooms. *Tourism Planning & Development, 10*(1), 32–44.

Njerekai, C. (2019). Hotel characteristics and the adoption of demand oriented hotel green practices in Zimbabwe: A regression. *African Journal of Hospitality, Tourism and Leisure, 8*(2), 1–16.

Noome, K., & Fitchett, J.M. (2019). An assessment of the climatic suitability of Afriski Mountain Resort for outdoor tourism using the Tourism Climate Index. *Journal of Mountain Science, 16*(11), 2453–2469.

Noome, K., & Fitchett, J.M. (2021). Quantifying the climate suitability for tourism in Namibia using the Tourism Climatic Index (TCI). *Environment, Development and Sustainability*. https://doi.org/10.1007/s10668-021-01651-2

Nowak, J.J., Petit, S., & Sahli, M. (2010). Tourism and globalization: The international division of tourism production. *Journal of Travel Research, 49*(2), 228–245.

Nunkoo, R. (2017). Governance and sustainable tourism: What is the role of trust, power and social capital? *Journal of Destination Marketing & Management, 6*, 277–285.

Nxumalo, T. (2017). *The influence of service quality of the post-dining behavioural intentions of customers cargo hold, Ushaka Marine World*. Master's Dissertation submitted to Durban University of Technology.

Nyaupane, G.P., & Chhetri, N. (2009). Vulnerability to climate change of nature-based tourism in the Nepalese Himalayas. *Tourism Geographies, 11*(1), 95–119.

Odeku, K.O. (2008). Sustainable protection and preservation of heritage sites attractions from climate change in South Africa. *African Journal of Hospitality, Tourism and Leisure, 7*(3), 1–12.

OECD. (2018). *Climate-resilient infrastructure. Policy perspectives. OECD Environment Policy Paper No.* 14. OECD.

O'Laughlin, B. (2002). Proletarianisation, agency and changing rural livelihoods: Forced labour and resistance in Colonial Mozambique. *Journal of Southern African Studies, 28*(3), 511–530.

Olsson, P., Folke, C., & Berkes, F. (2004). Adaptive comanagement for building resilience in socio-ecological systems. *Environmental Management, 34*(1), 75–90.

Olya, H.G., Alipour, H., Peyravi, B., & Dalir, S. (2019). Tourism climate insurance: Implications and prospects. *Asia Pacific Journal of Tourism Research, 24*(4), 269–280.

Onslow, S. (2011). *Zimbabwe and political transition.* Strategic Update. The London School of Economics and Political Science.

Oreskes, N. (2004). The scientific consensus on climate change. *Science, 306*(5702), 1686–1686.

Otto, F.E., Wolski, P., Lehner, F., Tebaldi, C., Van Oldenborgh, G.J., Hogesteeger, S., Singh, R., Holden, P., Fučkar, N.S., Odoulami, R.C., & New, M. (2018). Anthropogenic influence on the drivers of the Western Cape drought 2015–2017. *Environmental Research Letters, 13*(12), aae9f9.

Palanisamy, H., Cazenave, A., Meyssignac, B., Soudarin, L., Wöppelmann, G., & Becker, M. (2014). Regional sea level variability, total relative sea level rise and its impacts on islands and coastal zones of Indian Ocean over the last sixty years. *Global and Planetary Change, 116*, 54–67.

Palgan, Y.V., Zvolska, L., & Mont, O. (2017). Sustainability framings of accommodation sharing. *Environmental Innovation and Societal Transitions, 23*, 70–83.

Pandy, W.R. (2017). Tourism enterprises and climate change: Some research imperatives. *African Journal of Hospitality, Tourism and Leisure, 6*(4), 1–18.

Pandy, W.R., & Rogerson, C.M. (2018). Tourism and climate change: Stakeholder perceptions of at risk tourism segments in South Africa. *Euroeconomica, 37*(2), 104–118.

Pandy, W.R., & Rogerson, C.M. (2019). Urban tourism and climate change: Risk perceptions of business tourism stakeholders in Johannesburg, South Africa. *Urbani Izziv, 30*, 225–243.

Pandy, W.R., & Rogerson, C.M. (2020). Tourism industry perspectives on climate change in South Africa. In J.M. Rogerson, & G. Visser (Eds.), *New directions in South African tourism geographies* (pp. 93–111). Springer.

Pandy, W.R., & Rogerson, C.M. (2021). Climate change risks and tourism in South Africa: Projections and policy. *GeoJournal of Tourism and Geosites, 35*(2), 445–455.

Pascale, S., Kapnick, S.B., Delworth, T.L., & Cooke, W.F. (2020). Increasing risk of another Cape Town "Day Zero" drought in the 21st century. *Proceedings of the National Academy of Sciences, 117*(47), 29495–29503.

Pascale, S., Lucarini, V., Feng, X., Porporato, A., & Hasson, S. (2016). Projected changes of rainfall seasonality and dry spells in a high greenhouse gas emission scenario. *Climate Dynamics, 46*, 1331–1350.

Paul, A.H. (1972). Weather and the daily use of outdoor recreation areas in Canada. In J. Taylor (Ed.), *Weather forecasting for agriculture and industry* (pp. 132–146). David and Charles Publishers.

Peduzzi, P., Chatenoux, B., Dao, H., De Bono, A., Herold, C., Kossin, J., Mouton, F., & Nordbeck, O. (2012). Global trends in tropical cyclone risk. *Nature Climate Change, 2*(4), 289–294.

Peeters, O., Higham, J., Cohen, S., Eijgelaar, E., & Gössling, S. (2019). Desirable tourism transport futures. *Journal of Sustainable Tourism, 27*(2), 173–188.

Peeters, P., & Dubois, G. (2010). Tourism travel under climate change mitigation constraints. *Journal of Transport Geography, 18,* 447–457.Perch-Nielsen, S.L. (2010). The vulnerability of beach tourism to climate change – An index approach. *Climatic Change, 100,* 579–606.

Perch-Nielsen, S.L., Amelung, B., & Knutti, R. (2010). Future climate resources for tourism in Europe based on the daily Tourism Climatic Index. *Climatic Change, 103*(3–4), 363–381.

Perry, A. (2006). Will predicted climate change compromise the sustainability of Mediterranean tourism? *Journal of Sustainable Tourism, 14*(4), 367–375.

Pienaar, N., Dyson, L.L., & Klopper, E. (2015). Atmospheric anomalies during the 2012/13 extreme hail seasons over Gauteng. *Proceedings of the 31st Annual conference of South African society for atmospheric sciences,* September 2015, 149–152.

Pillay, M.T., & Fitchett, J.M. (2019). Tropical cyclone landfalls south of the Tropic of Capricorn, southwest Indian Ocean. *Climate Research, 79*(1), 23–37.

Pillay, M.T., & Fitchett, J.M. (2021). Southern hemisphere tropical cyclones: A critical analysis of regional characteristics. *International Journal of Climatology, 41*(1), 146–161.

Pillay, S. (2019). *The climate sensitivity of tourists visiting South Africa: An analysis of TripAdvisor reviews.* eScience MSc Capstone Project submitted to the University of the Witwatersrand.

Plummer, R., & Fennell, D.A. (2009). Managing protected areas for sustainable tourism: Prospects for adaptive co-management. *Journal of Sustainable Tourism, 17*(2), 149–168.

Poon, J.P.H., & Cheon, P. (2009). Objectivity, subjectivity, and intersubjectivity in economic geography: Evidence from the internet and blogosphere. *Annals of the Association of American Geographers, 99*(3), 590–603.

Presidential Task Force. (2016). *Vision 2036: Achieving prosperity for all.* Lentswe La Lesedi Pty (Ltd).

Preston-Whyte, R. (2001). Constructed leisure space: The seaside at Durban. *Annals of Tourism Research, 28*(3), 581–596.

Preston-Whyte, R., & Scott, D. (2007). Urban tourism in Durban. In C.M. Rogerson, & G. Visser (Eds.), *Urban tourism in the developing world: The South African experience* (pp. 245–264). Transaction Publishers.

Preston-Whyte, R.A., & Watson, H.K. (2005). Nature tourism and climatic change in southern Africa. In C.M. Hall, & J. Higham (Eds.), *Tourism, recreation and climate change* (pp. 130–142). Channelview Publications.

Prideaux, B., Thompson, M., & Pabel, A. (2020). Lessons from COVID-19 can prepare global tourism for the economic transformation needed to combat climate change. *Tourism Geographies, 22*(3), 667–78.

Prinsloo, A.S. (2019). *A case study of resource consumption in the sharing economy: Airbnb as tourist accommodation in Cape Town, South Africa.* Masters dissertation submitted to the University of the Witwatersrand, Johannesburg.

Ragoonaden, S. (1997). Impact of sea-level rise on Mauritius. *Journal of Coastal Research, 24,* 205–223.

Ragoonaden, S., Seewoobaduth, J., & Cheenacunnan, I. (2017). Recent acceleration of sea level rise in Mauritius and Rodrigues. *Western Indian Ocean Journal of Marine Science, 1,* 51–65.

Ramutsindela, M., & Mickler, D. (2020). Global goals and African development. In M. Ramutsindela, & D. Mickler (Eds.), *Africa and the sustainable development goals* (pp. 1–9). Springer.

Rapolaki, R.S., Blamey, R.C., Hermes, J.C., & Reason, C.J.C. (2019). A classification of synoptic weather patterns linked to extreme rainfall over the Limpopo River Basin in southern Africa. *Climate Dynamics, 53*, 2265–2279.

Reason, C.J.C., & Smart, S. (2015). Tropical south east Atlantic warm events and associated rainfall anomalies over southern Africa. *Frontiers in Environmental Science, 3*(24), 1–11.

Republic of South Africa. (2019). *National climate change adaptation strategy.* Department: Environment, Forestry and Fisheries.

Republic of Zimbabwe. (2017). *National climate policy.* Republic of Zimbabwe, Ministry of Environment, Water and Climate.

Rhodes, R.A.W. (1996). The new governance: Governing without government. *Political Studies, 44*, 652–667.

Rhodes, R.A.W. (1997). *Understanding governance: Policy, networks, governance, reflexivity and accountability.* Open University Press.

Richardson, L. (2015). Performing the sharing economy. *Geoforum, 67*, 121–129.

Richie, B.W., Kemperman, A., & Dolnicar, S. (2021). Which type of product attributes lead to aviation voluntary carbon offsetting among air passengers? *Tourism Management, 85*, 104276.

Rico, A., Martínez-Blanco, J., Montlleó, M., Rodríguez, G., Tavares, N., Arias, A., & Oliver-Solà, J. (2019). Carbon footprint of tourism in Barcelona. *Tourism Management, 70*, 491–504.

Rink, B. (2017). The aeromobile tourist gaze: Understanding tourism 'from above'. *Tourism Geographies, 19*(5), 878–896.

Rink, B. (2020). Cruising nowhere: A South African contribution to cruise tourism. In J. Rogerson, & G. Visser (Eds.), *New directions in South African tourism geographies* (pp. 249–266). Springer Nature.

Ritchie, H., & Roser, M. (2018). *Urbanization.* https://ourworldindata.org/urbanization

Robbins, P. (2012). *Political ecology: A critical introduction.* Wiley-Blackwell.

Roberts, D. (2010). Prioritizing climate change adaptation and local level resilience in Durban, South Africa. *Environment and Urbanization, 22*(2), 397–413.

Roberts, D., Douwes, J., Sutherland, C., & Sim, V. (2020). Durban's 100 resilient cities journey: Governing resilience from within. *Environment and Urbanization, 32*(2), 547–568.

Robins, S. (2019). 'Day Zero', hydraulic citizenship and the defence of the commons in Cape Town: A case study of the politics of water and its infrastructures (2017–2018). *Journal of Southern African Studies, 45*(1), 5–29.

Robinson, A.-S. (2020). Climate change adaptation in SIDS: A systematic review of the literature pre and post the IPCC Fifth Assessment Report. *WIREs Climate Change, 11*(4), e653.

Robinson, M., & Shine, T. (2018). Achieving a climate justice pathway to 1.5°C. *Nature Climate Change, 8*, 564–569.

Rockström, J., & Klum, M. (2012). *The human quest – Prospering within planetary boundaries.* Bokförlaget Langenskiöld.

Roe, D., Nelson, F., & Sandbrook, C. (2009). *Community management of natural resources in Africa: Impacts, experiences and future directions.* International Institute for Environment and Development.

Roffe, S.J., Fitchett, J.M., & Curtis, C.J. (2019). Classifying and mapping rainfall seasonality in South Africa: A review. *South African Geographical Journal, 101*(2), 158–174.

Roffe, S.J., Fitchett, J.M., & Curtis, C.J. (2020). Determining the utility of a percentile-based wet-season start-and end-date metrics across South Africa. *Theoretical and Applied Climatology, 140*, 1331–1347.

Roffe, S.J., Fitchett, J.M., & Curtis, C.J. (2021). Investigating changes in rainfall seasonality across South Africa: 1987–2016. *International Journal of Climatology, 41*(S1), 2031–2050.

Rogerson, C.M. (2011). Urban tourism and regional tourists: Shopping in Johannesburg, South Africa. *Tijdschrift voor Economische en Sociale Geografie, 102*(3), 316–330.

Rogerson, C.M. (2013). Urban tourism, economic regeneration and inclusion: Evidence from South Africa. *Local Economy, 28*(2), 188–202.

Rogerson, C.M. (2016). Climate change, tourism and local economic development in South Africa. *Local Economy, 31*(1–2), 322–331.

Rogerson, C.M. (2017). Unpacking directions and spatial patterns of VFR travel mobilities in the Global South: Insights from South Africa. *International Journal of Tourism Research, 19*, 466–475.

Rogerson, C.M., & Baum, T. (2020). COVID-19 and African tourism research agendas. *Development Southern Africa, 37*(5), 727–741.

Rogerson, C.M., & Hoogendoorn, G. (2014). VFR travel and second home tourism: The missing link? The case of South Africa. *Tourism Review International, 18*(3), 167–178.

Rogerson, C.M., & Lisa, Z. (2005). 'Sho't left': Changing domestic tourism in South Africa. *Urban Forum, 16*, 88–11.

Rogerson, C.M., & Rogerson, J.M. (2011). Tourism research within the southern African Development Community: Production and consumption of in academic journals, 2000–2010. *Tourism Review International, 15*, 213–224.

Rogerson, C.M., & Rogerson, J.M. (2020). COVID-19 and tourism spaces of vulnerability in South Africa. *African Journal of Hospitality, Tourism and Leisure, 9*(4), 382–401.

Rogerson, C.M., & Saarinen, J. (2018). Tourism for poverty alleviation: Issues and debates in the Global South. In C. Cooper, S. Volo, W.C. Gartner, & N. Scott (Eds.), *The SAGE handbook of tourism management: Applications of theories and concepts to tourism* (pp. 22–37). SAGE Publications Ltd.

Rogerson, C.M., & van der Merwe, C.D. (2016). Heritage tourism in the global South: Development impacts of the Cradle of Humankind World Heritage Site, South Africa. *Local Economy, 31*(1–2), 234–248.

Rogerson, C.M., & Visser, G. (2007). *Urban tourism in the developing world: The South African experience.* Transaction Publishers.

Rogerson, J.M., & Sims, S.R. (2012). The greening of urban hotels in South Africa: Evidence from Gauteng. *Urban Forum, 23*(3), 391–407.Rogerson, J.M., & Visser, G. (Eds.) (2020). *New directions in South African tourism geographies.* Springer.

Rogerson, J.M., & Wolfaardt, Z. (2015). Wedding tourism in South Africa: An exploratory analysis. *African Journal of Hospitality, Tourism and Leisure, 4*(2), 1–13.

Roheemun, A.B.A. (2018). *Climate change adaptation and mitigation strategies: The case of the hotel industry in Mauritius* [Unpublished bachelor's thesis]. University of Mauritius.

Rohli, R.V., Joyner, T.A., Reynolds, S.J., Shaw, C., & Vázquez, J.R. (2015). Globally extended Köppen-Geiger climate classification and temporal shifts in terrestrial climate types. *Journal of Physical Geography, 36*(2), 142–157.

Rosenzweig, C., Solecki, W.D., Blake, R., Bowman, M., Faris, C., Gornitz, V., Horton, R., Jacob, K., LeBlanc, A., Leichenko, R., Linkin, M., Major, D., O'Grady, M., Patrick, L., Sussman, E., Yohe, G., & Zimmerman, R. (2011). Developing coastal adaptation to climate change in the New York City infrastructure-shed: Process, approach, tools, and strategies. *Climatic Change, 106*, 92–127.

Roshan, G., Yousefi, R., & Fitchett, J.M. (2016). Long-term trends in tourism climate index scores for 40 stations across Iran: the role of climate change and influence on tourism sustainability. *International Journal of Biometeorology, 60*(1), 33–52.

Rubel, F., & Kottek, M. (2010). Observed and projected climate shifts 1901–2100 depicted by world maps of the Köppen-Geiger climate classification. *Meteorologische Zeitschrift, 19*(2), 135–141.

Ruhanen, L., Scott, N., Brent, R., & Tkaczynski, A. (2010). Governance: A review and synthesis of the literature. *Tourism Review, 65*, 4–16.

Rutty, M., & Scott, D. (2010). Will the Mediterranean become "too hot" for tourism? A reassessment. *Tourism and Hospitality Planning & Development, 7*(3), 267–281.

Rutty, M., Scott, D., Matthews, L., Burrowes, R., Trotman, A., Mahon, R., & Charles, A. (2020). An inter-comparison of the Holiday Climate Index (HCI: Beach) and the Tourism Climate Index (TCI) to explain Canadian tourism arrivals to the Caribbean. *Atmosphere, 11*(4), 412.

Rutty, M., Steiger, R., Demiroglu, O.C., & Perkins, D.R. (2021). Tourism climatology: Past, present, and future. *International Journal of Biometeorology, 65*, 639–643.

Saarinen, J. (2016). Political ecologies and economies of tourism development in Kaokoland, North-West Namibia. In M. Mostafanezhad, A. Carr, & R. Norum (Eds.), *Political ecology of tourism: Communities, power and the environment* (pp. 213–230). Routledge.

Saarinen, J. (2018). Beyond growth thinking: The need to revisit sustainable development in tourism. *Tourism Geographies, 20*(2), 337–340.

Saarinen, J. (2019). Communities and sustainable tourism development: Community impacts and local benefit creation tourism. In S.F. McCool, & K. Bosak (Eds.), *A research agenda for sustainable tourism* (pp. 206–222). Edward Elgar Publishing.

Saarinen, J. (2020). Tourism and sustainable development goals: Research on sustainable tourism geographies. In J. Saarinen (Ed.), *Tourism and sustainable development goals: Research on sustainable tourism geographies* (pp. 1–10). Routledge.

Saarinen, J. (2021a). Is being responsible sustainable in tourism? Connections and critical differences. *Sustainability, 13*, 6599.

Saarinen, J. (2021b). Sustainable growth in tourism? Rethinking and resetting sustainable tourism for development. In C.M. Hall, L. Lundmark, & J. Zhang (Eds.), *Degrowth and tourism: New perspectives on tourism entrepreneurship, destinations and policy* (pp. 135–151). Routledge.

Saarinen, J., Becker, F., Manwa, H., & Wilson, D. (Eds.) (2009). *Sustainable tourism in southern Africa: Local communities and natural resources in transition.* Channelview Publishers.

Saarinen, J., & Gill, A.M. (2019). Tourism, resilience and governance strategies in the transition towards sustainability. In J. Saarinen, & A.M. Gill (Eds.), *Resilient destinations: Governance strategies in the transition towards sustainability in tourism* (pp. 15–33). Routledge.

Saarinen, J., Hambira, W.L., Atlhopheng, J., & Manwa, H. (2012). Tourism industry reaction to climate change in Kgalagadi South District, Botswana. *Development Southern Africa, 29*(2), 273–285.

Saarinen, J., Moswete, N., Atlhopheng, J., & Hambira, W.L. (2020). Changing socio-ecologies of Kalahari: Local perceptions towards environmental change and tourism in Kgalagadi, Botswana. *Development Southern Africa, 37*(5), 855–870.

Saarinen, J., Moswete, N., & Monare, M.J. (2014). Cultural tourism: New opportunities for diversifying the tourism industry in Botswana. *Bulletin of Geography: Socio–Economic Series, 26*, 7–18.

Saarinen, J., Rogerson, C., & Hall, C.M. (2017). Geographies of tourism development and planning. *Tourism Geographies, 19*(3), 307–317.

Saarinen, J., & Rogerson, J.M. (2021). Tourism and change: Issues and challenges in the Global South. In J. Saarinen, & J.M. Rogerson (Eds.), *Tourism, change and the Global South* (pp. 3–14). Routledge.

Saayman, A., & Saayman, M. (2008). Determinants of inbound tourism to South Africa. *Tourism Economics, 14*(1), 81–96.

Saayman, M., Rossouw, R., & Krugell, W. (2012). The impact of tourism in poverty in South Africa. *Development Southern Africa, 29*(3), 462–487.

SADC (Southern African Development Community). (2021). *Southern African Development Community: Towards a common future.* https://www.sadc.int/about-sadc/overview/

Sailunaz, K., & Alhajj, R. (2019). Emotion and sentiment analysis from Twitter text. *Journal of Computational Science, 36*, 101003.

SANS204. (2011). *Energy efficiency in buildings.* SABS Standards Division.

Santarém, F., Saarinen, J., & Brito, J.C. (2020). Mapping and analysing cultural ecosystem services in conflict areas. *Ecological Indicators, 110*, 105943.

Santarém, F., Saarinen, J., & Brito, J.C. (2021, July 10). Assessment and prioritisation of cultural ecosystem services in the Sahara-Sahelian region. *Science of the Total Environment, 777*, 146053.

Sarmento, J., & Rink, B. (2016). Africa. In J. Jafari, & H. Xiao (Eds.), *Encyclopedia of tourism.* Springer.

SATSA. (2019). SATSA and SA tourism lead carbon offsetting initiative. https://www.satsa.com/satsa-and-sa-tourism-lead-carbon-offsetting-initiative/

Saunders, C. (2009). Namibian solidarity: British support for Namibian independence. *Journal of Southern African Studies, 35*(2), 437–454.

Scerri, E.M., Kühnert, D., Blinkhorn, J., Groucutt, H.S., Roberts, P., Nicoll, K., Zerboni, A., Orijemie, E.A., Barton, H., Candy, I., & Goldstein, S.T. (2020). Field-based sciences must transform in response to COVID-19. *Nature Ecology & Evolution, 4*, 1571–1574.

Scheyvens, R. (2018). Linking tourism to the sustainable development goals: A geographical perspective. *Tourism Geographies, 20*(2), 341–342.

Schliephack, J., & Dickinson, J.E. (2017). Tourists' representations of coastal managed realignment as a climate change adaptation strategy. *Tourism Management, 59*, 182–192.

Schlosberg, D., & Collins, L.B. (2014). From environmental to climate justice: Climate change and the discourse of environmental justice. *WIREs Climate Change, 5*, 359–374.

Schneider, T. (2014). Responsibility for private sector adaptation to climate change. *Ecology and Society, 19*(2), 8.

Scott, D. (2021). Sustainable tourism and the grand challenge of climate change. *Sustainability, 13*, 1–17.

Scott, D., Gössling, S., & Hall, C.M. (2012). International tourism and climate change. *WIREs Climate Change, 3*(3), 213–232.

Scott, D., Gössling, S., Hall, C.M., & Peeters, P. (2016a). Can tourism be part of the decarbonized global economy? The costs and risks of alternate carbon reduction policy pathways. *Journal of Sustainable Tourism, 24*, 52–72.

Scott, D., Hall, C.M., & Gössling, S. (2012). *Tourism and climate change: Impacts, adaptation and mitigation*. Routledge.

Scott, D., Hall, C.M., & Gössling, S. (2016b). A review of the IPCC fifth assessment and implications for tourism sector climate resilience and decarbonization. *Journal of Sustainable Tourism, 24*(1), 8–30.

Scott, D., Hall, C.M., & Gössling, S. (2016c). A report on the Paris climate change agreement and its implications for tourism: why we will always have Paris. *Journal of Sustainable Tourism, 24*(7), 933–948.

Scott, D., Hall, C.M., & Gössling, S. (2019). Global tourism vulnerability to climate change. *Annals of Tourism Research, 77*, 49–61.

Scott, D., & McBoyle, G. (2001). Using a modified "tourism climate index" to examine the implications of climate change for climate as a natural resource for tourism. In A. Matzarakis, & C.R. De Freitas (Eds.), *Proceedings of the first international workshop on climate, tourism and recreation*, Halkidiki, 5–10 December 2001, 69–88.

Scott, D., McBoyle, G., Minogue, A., & Mills, B. (2006). Climate change and the sustainability of ski-based tourism in eastern north America: A reassessment. *Journal of Sustainable Tourism, 14*(4), 376–398.

Scott, D., Peeters, P., & Gössling, S. (2010). Can tourism deliver its "aspirational" greenhouse gas emission reduction targets? *Journal of Sustainable Tourism, 18*(3), 393–408.

Scott, D., Rutty, M., Amelung, B., & Tang, M. (2016d). An inter-comparison of the holiday climate index (HCI) and the tourism climate index (TCI) in Europe. *Atmosphere, 7*(6), 80.

Scott, D., Rutty, M., & Peister, C. (2018). Climate variability and water use on golf courses: Optimization opportunities for a warmer future. *Journal of Sustainable Tourism, 26*(8), 1453–1467.

Scott, D., Wall, G., & McBoyle, G. (2005). The evolution of climate change issue in the tourism sector. In C.M. Hall, & J. Higham (Eds.), *Tourism, recreation and climate change* (pp. 44–60). Channel View Press.

Senapathi, D., Underwood, F., Black, E., Nicoll, M.A., & Norris, K. (2010). Evidence for long-term regional changes in precipitation on the East Coast Mountains in Mauritius. *International Journal of Climatology, 30*(8), 1164–1177.

Seneviratne, S.I., Zhang, X., Adnan, M., Badi, W., Dereczynski, C., Di·Luca, A., Ghosh, S., Iskandar, I., Kossin, J., Lewis, S., Otto, F., Pinto, I., Satoh, M., Vicente-Serrano, S.M., Wehner, M., & Zhou, B. (2021). Weather and climate extreme events in a changing climate. In *Climate change 2021: The physical science basis. Contribution of Working Group I to the sixth assessment report of the intergovernmental panel on climate change*. Cambridge University Press (in press).

Sepula, M.B., & Korir, J.C. (2019). Tourism and hospitality policy strengths, weaknesses and effectiveness–lessons for Malawi. *International Journal of Economics, Business and Management Research, 3*(5), 140–154.

Serdeczny, O., Adams, S., Baarsch, F., Coumou, D., Robinson, A., Hare, W., Schaeffer, M., Perrette, M., & Reinhardt, J. (2016). Climate change impacts in Sub-Saharan Africa: From physical changes to their social repercussions. *Regional Environmental Change, 17*(6), 1585–1600.

Shackleton, S., Ziervogel, G., Sallu, S., Gill, T., & Tschakert, P. (2015). Why is socially-just climate change adaptation in sub-Saharan Africa so challenging? A review of barriers identified from empirical cases. *WIREs Climate Change, 6*, 321–344.

Shakouri, B., Yadzi S.K., & Ghorchebigi, E. (2017). Does tourism development promote CO_2 emissions? *Anatolia, 28*(3), 444–452.

Sharpley, R. (2000). Tourism and sustainable development: Exploring the theoretical divide. *Journal of Sustainable Tourism, 8*(1), 1–19.

Sheller, M., & Urry, J. (2004). *Tourism mobilities: Places to play, places in play.* Routledge.

Shepherd, N. (2021). Cape Town's "Day Zero" drought: Notes on a future history of urban dwelling. *Space and Culture.* https://doi.org/10.1177/1206331221997695

Siakwah, P., Musavengane, R., & Leonard, L. (2020). Tourism governance and attainment of the sustainable development goals in Africa. *Tourism Planning & Development, 17*(4), 355–383.

Sibanda, S., Grab, S.W., & Ahmed, F. (2018). Spatio-temporal temperature trends and extreme hydro-climatic events in southern Zimbabwe. *South African Geographical Journal, 100*(2), 210–232.

Sigala, M. (2020). Tourism and COVID-19: Impacts and implications for advancing and resetting industry and research. *Journal of Business Research, 117*, 312–321.

Simmie, J., & Martin, R. (2010). The economic resilience of regions: Towards an evolutionary approach. *Cambridge Journal of Regions, Economy and Society, 3*(1), 27–43

Singleton, A.T., & Reason, C.J.C. (2007). Variability in the characteristics of cut-off low pressure systems over subtropical southern Africa. *International Journal of Climatology, 27*(3), 295–310.

Slocum, S., & Kline, C. (2014). Regional resilience: Opportunities, challenges and policy messages from Western North Caroline. *Anatolia, 25*(3), 403–416.

Smit, B., & Pilifosova, O. (2001). Adaptation to climate change in the context of sustainable development and equity. In J.J. McCarthy, O. Canziani, N.A. Leary, D.J. Dokken, & K.S. White (Eds.), *Climate change 2001: Impacts, adaptation and vulnerability. IPCC working group II* (pp. 877–912). Cambridge University Press.

Smit, B., & Wandel, J. (2006). Adaptation, adaptive capacity and vulnerability. *Global Environmental Change, 16*, 282–292.

Smit, I.P., Smit, C.F., Govender, N., Linde, M.V.D., & MacFadyen, S. (2013). Rainfall, geology and landscape position generate large-scale spatiotemporal fire pattern heterogeneity in an African savanna. *Ecography, 36*(4), 447–459.

Smith, B.C. (2007). *Good governance and development.* Palgrave.

Smith, K. (1990). Tourism and climate change. *Land Use Policy, 7*(2), 176–180.

Smith, P. (1998). *The history of tourism: Thomas Cook and the origins of leisure travel.* Routledge.

Smith, T., & Fitchett, J.M. (2020). Drought challenges for nature tourism in the Sabi Sands Game Reserve in the eastern region of South Africa. *African Journal of Range and Forage Science, 37*(1), 107–117.

Snyman, P.L. (2020). *Phenological advance in the South African Namaqualand Daisy Bloom over the past decades.* Masters Dissertation submitted to the University of the Witwatersrand, Johannesburg.

Song, J., Klotzbach, P.J., Tang, J., & Wang, Y. (2018). The increasing variability of tropical cyclone lifetime maximum intensity. *Scientific Reports, 8*(1), 1–7.

Sousa, P.M., Blamey, R.C., Reason, C.J.C., Ramos, A.M., & Trigo, R.M. (2018). The 'Day Zero' Cape Town drought and the poleward migration of moisture corridors. *Environmental Research Letters, 13*, aaebc7.

South Africa Tourism. (2021). South African tourism dashboard. https://www.southafrica.net/gl/en/corporate/page/international-tourist-arrivals-report

South Africa's Implementation of the 2030 Agenda for Sustainable Development. (2019). Voluntary national review (VNR) report 2019. https://sustainabledevelopment.un.org/content/documents/23402SOUTH_AFRICA_RSA_Voluntary_National_Review_Report_Final__14_June_2019.pdf

Southon, M.P., & van der Merwe, C.D. (2018). Flooded with risks or opportunities: Exploring flooding impacts on tourist accommodation. *African Journal of Hospitality, Tourism and Leisure, 7*(1), 1–16.

Spear, D., Zaroug, M.A., Daron, J.D., Ziervogel, G., Angula, M.N., Haimbili, E.N., & Davies, J.E. (2018). *Vulnerability and responses to climate change in drylands: The case of Namibia. CARIAA-ASSAR Working Paper.* CARIAA.

Spenceley, A. (2005). Nature-based tourism and environmental sustainability in South Africa. *Journal of Sustainable Tourism, 13*(2), 136–170.

Spooner, S. (2014). *8 things you probably didn't know about Lesotho.* https://mg.co.za/article/2014-09-01-unusual-coup-in-africas-unusual-kingdom-in-the-sky-7-things-you-didnt-know-about-lesotho

Stander, J.H., Dyson, L., & Engelbrecht, C.J. (2016). A snow forecasting decision tree for significant snowfall over the interior of South Africa. *South African Journal of Science, 112*(9–10), 1–10.

Statistics Botswana. (2020). Tourism statistics annual report 2018. https://www.statsbots.org.bw/tourism-statistics-annual-report-2018

StatsSA. (2019). *Tourism. Report No. 03–51–02.* Statistics South Africa.

StatsSA. (2020). *Tourism and migration. Statistical release P0351.* STATS SA, Pretoria.

Steiger, R., & Scott, D. (2020). Ski tourism in a warmer world: Increased adaptation and regional economic impacts in Austria. *Tourism Management, 77*, 104032.

Steyn, J.N. (2012). Managing climate change impacts on tourism: Mitigating and adaptive strategies with special reference to the Western Cape Province of South Africa. *African Journal of Physical, Health Education, Recreation and Dance, 18*(3), 552–564.

Steyn, J.N., & Spencer, J.P. (2012). Climate change and tourism: Implications for South Africa. *African Journal for Physical, Health Education, Recreation and Dance, 18*(1), 1–19.

Stockholm Resilience Centre. (2015). *What is resilience? An introduction to socio-ecological research.* Stockholm University.

Stockigt, L., Hoogendoorn, G., Fitchett, J.M., & Saarinen, J. (2018). Climate sensitivity and snow-based tourism in Africa: An investigation of TripAdvisor reviews on Afriski, Lesotho. *Proceedings of the Biennial Conference of the Society of South African Geographers*, Bloemfontein (1–5 October 2018), 207–225.

Stockigt, L.J. (2019). *Climate change and snow-based tourism in Africa: The Case of Afriski, Lesotho*. MA Dissertation submitted to the University of Johannesburg, Johannesburg.

Stone, M., Lenao, M., & Moswete, N. (Eds.) (2020). *Natural resources, tourism and community livelihoods in southern Africa*. Routledge.

Strickland-Munro, J., Allison, H., & Moore, S. (2010). Using resilience concepts to investigate the impacts of protected area tourism on communities. *Annals of Tourism Research*, *37*(2), 499–519.

Su, B. (2011). The impact of passive design factors on house energy efficiency. *Architectural Science Review*, *54*(4), 270–276.

Sun, Y-Y., Cadarso, M.A., & Driml, S. (2020). Tourism carbon footprint inventories: A review of the environmentally extended input-output approach. *Annals of Tourism Research*, *82*, 102928.

Sun, Y-Y., & Drakeman, D. (2020). Measuring the carbon footprint of wine tourism and cellar door sales. *Journal of Cleaner Production*, *266*, 121937.

Sun, Y-Y., & Hsu, C-M. (2019). The decomposition analysis of tourism water footprint in Taiwan: Revealing decision-relevant information. *Journal of Travel Research*, *58*(4), 695–708.

Sundaram, J.K., & Chowdhury, A. (Eds.) (2012). *Is good governance good for development?* Bloomsbury.

Svoboda, M., Hayes, M., & Wood, D. (2012*). Standardized precipitation index user guide*. World Meteorological Organization Geneva.

Swart, R., & Raes, F. (2007). Making integration of adaptation and mitigation work: Mainstreaming into sustainable development policies? *Climate Policy*, *7*(4), 288–303

Swemmer, A.M., Bond, W.J., Donaldson, J., Hempson, G.P., Malherbe, J., & Smit, I.P.J. (2018). The ecology of drought – A workshop report. *South African Journal of Science*, *114*(9/10), 1–3.

Tang, C., Zhong, L., & Ng, P. (2017). Factors that influence the tourism industry's carbon emissions: A tourism area life-cycle model perspective. *Energy Policy*, *109*, 704–718.

Tegegne, W.A., Moyle, B.D., & Becken, S. (2018). A qualitative system dynamics approach to understanding destination image. *Journal of Destination Marketing and Management*, *8*, 14–22.

Tervo-Kankare, K., Kajan, E., & Saarinen, J. (2018a). Costs and benefits of environmental change: Tourism industry's responses in Arctic Finland. *Tourism Geographies*, *20*(2), 202–223.

Tervo-Kankare, K., Saarinen, J., Kimaro, M.E., & Moswete, N. (2018b). Nature-based tourism operators' responses to changing environment and climate in Uis, Namibia. *African Geographical Review*, *37*(3), 273–282.

Thistlethwaite, J., Minano, A., Blake, J.A., Henstra, D., & Scott, D. (2018). Application of re/insurance models to estimate increases in flood risk due to climate change. *Geoenvironmental Disasters*, *5*(1), 1–13.

Thuiller, W., Broennimann, O., Hughes, G., Alkemade, J.R.M., Midgley, G.F., & Corsi, F. (2006). Vulnerability of African mammals to anthropogenic climate change under conservative land transformation assumptions. *Global Change Biology*, *12*, 424–440.

Tompkins, E.L., & Adger, W.N. (2004). Does adaptive management of natural resources enhance resilience to climate change? *Ecology and Society, 9*(2), 28–41.

Trading Economics. (2020). Mozambique – International tourism, number of arrivals. https://tradingeconomics.com/mozambique/international-tourism-number-of-arrivals-wb-data.html

Trochim, W.M.K. (2008). *Research methods knowledge base.* Cengage Learning.

Turton, B.J., & Mutambirwa, C.C. (1996). Air transport services and the expansion of international tourism in Zimbabwe. *Tourism Management, 17*(6), 453–462.

Tyson, P.D., & Preston-Whyte, R.A. (2004). *The weather and climate of southern Africa.* Oxford University Press.

UNDP (United Nations Development Programme). (1997). *Governance for sustainable human development.* New York.

UNFCC (United Nations Framework Convention on Climate Change). (2019). *Impacts of climate change on sustainable development goals highlighted at high-level political forum.* United Nations. https://unfccc.int/news/impacts-of-climate-change-on-sustainable-development-goals-highlighted-at-high-level-political-forum.

Unganai, L.S. (1997). Surface temperature variation over Zimbabwe between 1897 and 1993. *Theoretical and Applied Climatology, 56*, 89–101.

Unger, N. (2011). Global climate impact of civil aviation for standard and desulfurized jet fuel, *Geophysical Research Letters, 38*(20), 1–6.United Nations. (2015a). *Transforming our world: The 2030 agenda for sustainable development. Resolution adopted by the general assembly on 25 September 2015.* United Nations.

United Nations. (2015b). *Paris Agreement.* United Nations. https://unfccc.int/sites/default/files/english_paris_agreement.pdf

United Nations. (2016). United Nations treaty collection chapter XXVII environment, 7.d Paris Agreement, Paris, 12 December 2015 (C.N.92.201). New York. https://treaties.un.org/doc/Publication/CN/2016/CN.92.2016-Eng.pdf

United Nations. (2018). *UN statistics country profile.* United Nations. https://country-profiles.unstatshub.org/.

United Nations. (2021). *UN Research Roadmap for the COVID-19 recovery leveraging the power of science for a more equitable, resilient and sustainable future.* https://www.un.org/en/pdfs/UNCOVID19ResearchRoadmap.pdf

United Nations Office for Disaster Risk Reduction. (2020). Disaster. https://www.undrr.org/terminology/disaster

UNWTO (World Tourism Organisation). (2003). *Climate change and tourism. Proceedings of the first international conference on climate change tourism, Djerba, 9–11 April.* UNWTO.

UNWTO (World Tourism Organisation). (2007). *Davos Declaration. Climate change and tourism responding to global challenges. Second international conference on climate change and tourism, Davos, Switzerland.* UNWTO.

UNWTO (World Tourism Organization). (2008). *International recommendations for tourism statistics 2008* (IRTS 2008). New York.

UNWTO, UNEP, & WMO (World Tourism Organisation, United Nations Environment Program and World Meteorological Organisation). (2008). *Climate change and tourism: Responding to global challenges.* UNTWO.

Vaismoradi, M., Turunen, H., & Bondas, T. (2013). Content analysis and thematic analysis: Implications for conducting a qualitative descriptive study. *Nursing and Health Sciences, 15*(3), 398–405.

Valdivia, A., Luzón, M.V., & Herrera, F. (2017). Sentiment analysis in Tripadvisor. *IEEE Intelligent Systems, 32*(4), 72–77.

Vale, C.G., & Brito, J.C. (2015). Desert-adapted species are vulnerable to climate change: Insights from the warmest region on Earth. *Global Ecology and Conservation, 4*, 369–379.

van der Duim, R., Meyer, D., Saarinen, J., & Zellmer, K. (Eds.) (2011). *New alliances for tourism, conservation and development in eastern and southern Africa.* Eburon.

van der Walt, A., & Fitchett, J. M. (2020). Statistical classification of South African seasonal divisions on the basis of daily temperature data. *South African Journal of Science, 116*(9/10), 1–9.

van der Walt, A.J., & Fitchett, J.M. (2021a). Exploring extreme warm temperature trends in South Africa: 1960–2016. *Theoretical and Applied Climatology, 143*(3), 1341–1360.

van Der Walt, A.J., & Fitchett, J.M. (2021b). Trend analysis of cold extremes in South Africa: 1960–2016. *International Journal of Climatology, 41*(3), 2060–2081.

van Der Walt, A.J., & Fitchett, J.M. (2021c). Extreme Temperature Events (ETEs) in South Africa: A review. *South African Geographical Journal.* http://doi.org/10.108 0/03736245.2021.1907219.

Vanos, J.K., Middel, A., McKercher, G.R., Kuras, E.R., & Ruddell, B.L. (2016). Hot playgrounds and children's health: a multiscale analysis of surface temperatures in Arizona, USA. *Landscape and Urban Planning, 146*, 29–42.

Van Schalkwyk, J.A. (2009). *The Makgabeng/Blouberg heritage tourism: Tourism plan Blouberg Municipality, Capricorn District Limpopo Province.* Rock Art Research Institute.

Vincent, K. (2007). Uncertainty in adaptive capacity and the importance of scale. *Global Environmental Change, 17*(1), 12–24.

Visser, G., Erasmus, I., & Miller, M. (2017). Airbnb: The emergence of a new accommodation type in Cape Town, South Africa. *Tourism Review International, 21*(2), 151–168.

Visser, G., & Hoogendoorn, G. (2011). Current paths in South African tourism research. *Tourism Review International, 15*(1–2), 5–20.

Wachinger, G., Renn, O., Begg, C., & Kuhlicke, C. (2013). The risk perception paradox – Implications for governance and communication of natural hazards. *Risk Analysis, 33*(6), 1049–1065.

Wall, G. (1998). Implications of global climate change for tourism and recreation in wetland areas. *Climatic Change, 40*, 371–379.

Wall, G., & Badke, C. (1994). Tourism and climate change: An international perspective. *Journal of Sustainable Tourism, 2*(4), 193–203.

Wall, G., Harrison, R., Kinnaird, V., McBoyle G., & Quinlan, C. (1986). The implications of climate change for camping in Ontario. *Journal Recreation Research Review, 13*, 50–60.

Wall, G., & Mathieson, A. (2006). *Tourism: Change, impacts, and opportunities.* Pearson Prentice-Hall.

Walters, T. (2016). Using thematic analysis in tourism research. *Tourism Analysis, 21*(1), 107–116.

Wan, Y.K.P. (2012). Governance of tourism planning in Macao. *Tourism Analysis, 17*, 357–369.

Watson, P.J. (2017). Acceleration in European Mean Sea Level? A new insight using improved tools. *Journal of Coastal Research, 33*(1), 23–38.

Weaver, D. (2011). Can sustainable tourism survive climate change? *Journal of Sustainable Tourism, 19,* 5–15.

Westskog, H., Hovelsrud, G.K., & Sundqvist, G. (2017). How to make local context matter in national advice: Towards adaptive comanagement in Norwegian climate adaptation. *Weather, Climate, and Society, 9*(2), 267–283.

Wilhelm, M. (2013). *Impact of climate change in Namibia: A case study of Omusati region.* MSc Dissertation submitted to Polytechnic of Namibia, Windhoek.

Wilson, J.R., Lomonico, S., Bradley, D., Sievanen, L., Dempsey, T., Bell, M., McAfee, S., Costello, C., Szuwalski, C., McGonigal, H., Fitzgerald, S., & Gleason, M. (2018). Adaptive co-management to achieve climate-ready fisheries. *Conservation Letters, 11*(6), e12452.

Wilson, S.P., & Verlis, K.M. (2017). The ugly face of tourism: Marine debris pollution linked to visitation in the southern Great Barrier Reef, Australia. *Marine Pollution Bulletin, 117,* 239–246.

Woods-Ballard, B., Kellagher, R., Martin, P., Jefferies, C., Bray, R., & Shaffer, P. (2007). *The SUDS manual.* CIRIA.

Wolski, P. (2018). How severe is Cape Town's "Day Zero" drought. *Significance, 15,* 24–27.

World Meteorological Organization [WMO]. (2017). *WMO Guidelines on the calculation of climate normal.* WMO.

World Meteorological Organization [WMO]. (2020). Tropical cyclones. https://public.wmo.int/en/our-mandate/focus-areas/natural-hazards-and-disaster-risk-reduction/tropical-cyclones.

Wray, M. (2015). Drivers of change in regional tourism governance: A case analysis of the influence of the New South Wales Government, Australia, 2007–2013. *Journal of Sustainable Tourism, 23*(7), 990–1010.

Wright, C.Y., Kapwata, T., du Preez, D.J., Wernecke, B., Garland, R.M., Nkosi, V., Landman, W.A., Dyson, L., & Norval, M. (2021). Major climate change-induced risks to human health in South Africa. *Environmental Research, 196,* 110973.

Wyss, R., Abegg, B., & Luthe, T. (2014). Perceptions of climate change in a tourism governance context. *Tourism Management Perspectives, 11,* 69–76.

Xulu, N.G., Chikoore, H., Bopape, M-J., & Nethengwe, N.S. (2020). Climatology of the Mascarene High and its influence on weather and climate over southern Africa. *Climate, 8*(7), 86.

Ye, M., Wu, J., Liu, W., He, X., & Wang, C. (2020). Dependence of tropical cyclone damage on maximum wind speed and socioeconomic factors. *Environmental Research Letters, 15*(9), 094061.

Zengeni, N., Zengeni, D.M.F., & Muzambi, S. (2013). Hoteliers' perceptions of the impacts of green tourism on hotel operating costs in Zimbabwe: The case of selected Harare hotels. *Australian Journal of Business and Management Research, 2*(11), 64–73.

Zeppel, H. (2012). Collaborative governance for low-carbon tourism: Climate change initiatives by Australian tourism agencies. *Current Issues in Tourism, 15*(7), 603–626,

Zepper, H., & Beaumont, N. (2014). Climate change and sustainable tourism: Carbon mitigation by environmentally certified tourism enterprises. *Tourism Review International, 17,* 161–177.

Zhang, C. (2013). Madden-Julian oscillation: Bridging weather and climate. *Bulletin of the American Meteorological Society, 94*(12), 1849–1870.

Zhang, J., & Zhang, Y. (2018). Carbon tax, tourism CO_2 emissions and economic welfare. *Annals of Tourism Research, 69*, 18–30.

Zhang, J-H., Zhang, Y., Zhou, J., Liu, Z-H., Zhang, H-L., & Tian, Q. (2017). Tourism water footprint: An empirical analysis of Mount Huangshan. *Asia Pacific Journal of Tourism Research, 22*(10), 1083–1098.

Zhang, N., Ren, R., Zhang, Q., & Zhang, T. (2020). Air pollution and tourism development: An interplay. *Annals of Tourism Research, 85*, 103032.

Zhou, J., Zhang, G., Lin, Y., & Li, Y. (2008). Coupling of thermal mass and natural ventilation in buildings. *Energy and Buildings, 40*(6), 979–986.

Zibanai, Z. (2018). The tourism sector: A bright light in Zimbabwe's depressed economic environment. *African Journal of Hospitality, Tourism and Leisure, 7*(1), 1–3.

Ziervogel, G. (2019). *Unpacking the Cape Town drought: Lessons learned. Report for cities report programme. Undertaken by the African centre for cities. Climate resilience paper for the national treasury.* Republic of South Africa.

Ziervogel, G., New, M., Archer van Gaarderen, E., Midgley, G., Taylor, A., Hamann, R., Stuart-Hill, S., Myers, J., & Warburton, M. (2014). Climate change impacts and adaptation in South Africa. *WIRES Climate Change, 6*, 605–620.

Ziervogel, G., & Parnell, S. (2014). Tackling barriers to climate change adaptation in South African coastal cities. In B. Glavovic, & G. Smith (Eds.), *Adapting to climate change. environmental hazards* (pp. 57–73). Springer.

Ziervogel, G., Shale, M., & Du, M. (2010). Climate change adaptation in a developing country context: The case of urban water supply in Cape Town. *Climate and Development, 2*, 94–110.

Zimbabwe Tourism Authority. (2018). *Tourism trends & statistics report.* Domestic Tourism & Strategic Research Division.

Index

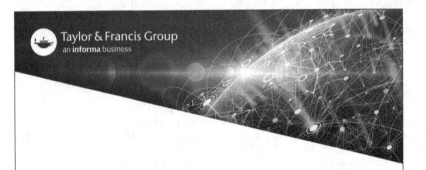

Printed in the United States
by Baker & Taylor Publisher Services